Electric Mountains

Nature, Society, and Culture

Scott Frickel, Series Editor

A sophisticated and wide-ranging sociological literature analyzing nature-society-culture interactions has blossomed in recent decades. This book series provides a platform for showcasing the best of that scholarship: carefully crafted empirical studies of socio-environmental change and the effects such change has on ecosystems, social institutions, historical processes, and cultural practices.

The series aims for topical and theoretical breadth. Anchored in sociological analyses of the environment, Nature, Society, and Culture is home to studies employing a range of disciplinary and interdisciplinary perspectives and investigating the pressing socio-environmental questions of our time—from environmental inequality and risk, to the science and politics of climate change and serial disaster, to the environmental causes and consequences of urbanization and war making, and beyond.

For a list of all the titles in the series, please see the last page of the book.

Electric Mountains

•••••••••••••••••••••••••••••

Climate, Power, and Justice in an Energy Transition

SHAUN A. GOLDING

Rutgers University Press

New Brunswick, Camden, and Newark, New Jersey, and London

Library of Congress Cataloging-in-Publication Data
Names: Golding, Shaun A., author.
Title: Electric mountains: climate, power, and justice in an energy
 transition / Shaun A. Golding.
Description: New Brunswick: Rutgers University Press, [2021] | Series: Nature, society, and
 culture | Includes bibliographical references and index.
Identifiers: LCCN 2020043873 | ISBN 9781978820685 (paperback) |
 ISBN 9781978820692 (cloth) | ISBN 9781978820708 (epub) |
 ISBN 9781978820715 (mobi) | ISBN 9781978820722 (pdf)
Subjects: LCSH: Energy policy—United States. | Renewable energy sources—New England. |
 Wind turbines—New England. | Sustainable development.
Classification: LCC HD9502.U5 G65 2021 | DDC 333.790973—dc23
LC record available at https://lccn.loc.gov/2020043873

A British Cataloging-in-Publication record for this book is available from the British Library.

♾ The paper used in this publication meets the requirements of the American National
Standard for Information Sciences—Permanence of Paper for Printed Library Materials,
ANSI Z39.48-1992.

www.rutgersuniversitypress.org

Manufactured in the United States of America

For my father, Philip

Contents

Illustrations

Figures

Table

Preface

I am sitting in my living room watching wet snow fall on Trondheim, Norway's downtown. The pale green steeple of the Nidaros Cathedral is barely visible. The fjord, and the mountains on its other side, are lost in fog. The snow is thick and sound-dampening and peaceful. If I were not "social distancing," this weather would thrill me. I love snow, and it has been another sad winter with little snow here on the coast. I am grateful for the few outings that I have had on my skis because I know that Oslo, where most Norwegians live, has seen virtually no snow. Recently, a group of Oslo residents trampled on their skis across bone-dry city streets to raise awareness of climate change, which made for sobering photos. But today's snow and clouds hang heavy over a city trying to slow the quickening spread of a virus that has already infected so much of the world population.

The weather reflects what people here in Norway seem to feel about COVID-19 early in the pandemic: half panic, half calm. The panic is from the novelty of it all. The calm comes from something distinctly Norwegian. The looming lockdown is uncertain for sure, but my colleague Johan Fredrik told me to expect the city to look as empty as Easter vacation. It is telling that what an American would describe as postapocalyptic, my Norwegian colleague describes as Easter. I've yet to experience Easter in Norway, but I know it is a lot like national quarantine. It comes when the days are getting longer, in that sliver of spring when the sun is high but snow lingers, or it used to linger. Most families isolate in the countryside. Schools, businesses, and stores close. There are few public events.

Obviously, mandatory quarantine will be different. Norway's citizens and institutions have a lot to figure out, and they are probably raising the same concerns as in other places, but Norwegian culture shows a fierce inclination for order. The call for isolation appeases thousands of Norwegians who had been pushing for aggressive measures to stem the exponential spread of the virus. I am sure that Fulbright scholars around the world are finding this sudden new reality intensely stressful, but I take heart from being in a country where shutting things down for a while is a normal and even celebrated practice. I wish more of the world felt OK hitting the pause button. The wet snow outside has just turned to rain.

The COVID-19 pandemic is the climate crisis, hypercompressed; some saw it coming well in advance, but leaders paid only lip service until they realized the threats were not all that distant. We can elect leaders that deny ecological problems until it is too late, spreading blame between their golf swings and tweets, or we can elect leaders who respect the time and energy that people invest to better understand our problems: the scientists, social workers, and community organizers. We have let powerful status quo institutions frame those people's passion as bias, when it is in fact our greatest asset.

Seeing other countries' response to the virus—the structured quarantines, mass testing, capacity to quickly build new hospitals, collaborative collection and analysis of health data—should instill quaking fear in the citizens of nations where public institutions have yielded so much capacity to neoliberal fantasies. Right now, America's profitable but fragmented and grossly unequal healthcare system is a train wreck the rest of the world is watching in slow motion. It turns out that thinking perpetual economic growth is the best way for a society to advance is absurd when that growth devours finite resources, and when care is only delivered when it is profitable. I do not want a totalitarian government to quarantine me in a containment facility, but it would be nice if the United States had a medical and public health system that functioned coherently and communicatively, just as it would be nice if we had climate policies that ratcheted down our greenhouse gas emissions systematically.

Climate change is a widely accepted reality in Northern New England. I want this book to help clarify why electricity policies in places with strong environmental concern can seem to deviate from the mission of climate sensibility even as they seem so clean and green. Like many others, I want my work to give context to energy debates and to promote more debates. With more people talking about electricity, we can change the relationship between humans and electrons and between humans and the environment more broadly.

COVID-19 will push our systems to their limits, but good can grow from bad. The pandemic has already created a dramatic, life-saving reduction in greenhouse gas emissions and airborne particulates, which makes it both poignant and politically grotesque that a deadly respiratory virus is creating the conditions for cleaner lungs and easier breathing. We need to ask ourselves how we got here. I am resigned to there being less snow, but maybe in a generation we will have more virologists, more climate scientists, and more brave and imaginative public officials who are able and willing to listen to them.

Shaun A. Golding
March 2020

Electric Mountains

1

Introduction

•••••••••••••••••••••

Election 2016

Proponents of renewable energy will remember November 8, 2016, as the day that American voters elected climate change denier Donald J. Trump the forty-fifth president of the United States. Trump's Electoral College victory promised to stunt eight years of energy policy that had propelled advancements in wind-generated electricity. However, the presidential election overshadowed a resounding defeat for renewable energy by voters in one of America's most progressive, liberal, and environmentally conscious states. Vermont, the home state of Bernie Sanders, the state that handed democratic nominee Hillary Clinton her largest margin of victory, elected a Republican governor who had campaigned on the promise of halting wind turbine development. It seems that while many watched with delight as wind energy took off in the state, others saw their delight turn to horror. Thus, as America's progressive voters and the national Democratic Party began to take stock of their messaging, platforms, and outreach efforts after the Trump election, proponents of wind energy in Vermont probably embarked on some reflecting of their own.

I intend for this book to aid proponents of renewable energy in this process of reflection. Vermont is in no way unique in its reversal on wind energy.

Communities throughout the world have organized in opposition to wind turbines. Still, the conflict around wind energy in Vermont and its Northern New England neighbors, Maine and New Hampshire, sheds light on a number of important social issues. The region's story of wind opposition encompasses the continued fracturing of the American environmental movement, struggles over natural resources in an economically depressed rural economy, and mounting tension between urban spaces of consumption and rural spaces of production.

Electricity prices in Northern New England rank among the highest in the nation, and thanks to a cultural inclination for thrift and a relatively temperate climate, the region's per capita electricity use ranks among the lowest. The region depends primarily on wood and oil for winter heat and has little need for air-conditioning in the summer months. Under these conditions, curtailing sprawl, extending public transit, and winterizing homes would be more effective and perhaps more politically popular ways to reduce carbon emissions than building wind turbines on remote, forested mountaintops. So why has the region so aggressively pursued wind turbine developments?

The short answer is that over the past decade, the will of the region's political machinery coalesced in unison with the technology and financial architecture required to make grid-scale wind turbines viable. Except for the Atlantic coast, mountains are the region's only places with sufficient wind to turn a profit. The longer answer unfolds through the next eight chapters. But much of what I will describe is crystallized in my first visit to see the wind turbines in Lowell, Vermont.

June 2013: Field Trip

On a clear and warm summer day in 2013, I mustered at the base of Lowell Mountain for a public tour of Vermont's newest, most controversial, and largest wind project. Twenty-one turbines carved into what was recently the state's largest patch of undisturbed wildlife habitat, the Kingdom Community Wind project (KCW) had touched off heated and well-publicized debate. In opposition to the project, local residents formed a group to strategize formal protests during construction. The group held organizing meetings, testified at public forums, wrote editorials, and actively maintained a website. When construction proceeded anyway, the group blockaded construction vehicles with their bodies and the blasting of the mountain with their lawyers.

Police arrested thirteen peaceful demonstrators, bringing the debate around wind energy to the front pages of newspapers across the state. In a state that fought passionately to close its nuclear plant, and where citizens, politics, and even businesses sometimes seem as green as the landscape, this response to renewable energy presented a puzzle. Like others viewing the wind project that day in 2013, I came to see the turbines for myself, to ask questions, and to attempt to reconcile the dramatic but seemingly necessary change to the landscape in my home state.

Before our tour began, we gathered for a question-and-answer session led by a representative of Green Mountain Power (GMP), KCW's co-owner and operator, and Vermont's largest monopoly electric utility. Our guide entertained us with KCW trivia, awarding Frisbees to the first person to answer basic questions about the project. Its size: twenty-one turbines. Acres of forest bulldozed for the project: 135. Acreage of permanent conservation easements purchased nearby: 2,800. Our guide directed our attention to a blade from one of the turbines lying on the ground nearby. It had been damaged on its journey to the site, and apparently the cost of returning it across hundreds of miles of roadways was prohibitive. The cavernous opening where the blade would connect to its base offered a dramatic visualization of a completed turbine's enormous size. Our tour guide devoted a few minutes to lauding turbines' capacity to generate electricity at reasonable costs—costs appreciably lower than other renewable technologies, he claimed.

Members of the crowd then posed several questions, many pushing for deeper explanations of the project's social and environmental impacts. One question inquired about the project's seemingly minimal impact on long-term employment for the region. GMP's representative explained that while construction jobs had ceased, maintenance crews were garnering expertise that created demand for their services locally and across the country. Another question lamented the turbines' twenty-one blinking, asynchronous red warning lights and asked why promises to make the lights activate only in the presence of aircraft had gone unfulfilled. The guide explained that the technology needed to operate such an automated system had been slow to develop but assured the crowd that it would eventually be installed.

After the question-and-answer session, we sorted ourselves into cars and ascended the ridgeline in a slow moving parade on the newly cut gravel road, weaving back and forth across the steep slope in a series of hairpin turns.

FIG 1.1 Lowell Mountain, Vermont, Kingdom Community Wind, June 2013 (Photo courtesy of Shaun A. Golding)

Reaching the summit, we were greeted by the enormous but somehow docile machines, white pillars framing a stunning view of Vermont's forest patchwork in the background. On that day, the turbines seemed quiet. The blades whirled overhead with a dull whoosh, casting rhythmic shadows across the flattened ridge, somehow both serene and spooky. Having hiked and skied in Vermont for my entire life, this gravel tabletop was new. But the turbines' subtle intrusion of sound and light, the industrial feel of a flattened mountain, and even the spraypaint marking birds killed by the turbines could not dilute splendor from the day's intense blue sky and the emerald landscape surrounding us.

This juxtaposition encapsulated the trade-off that energy production requires us to make as a society. One slightly defiled ridgeline may seem a small price to pay for energy that does not poison or pollute locally or threaten the climate globally. These are the trade-offs that proponents of Vermont's wind energy cite when describing their support. That renewable energy could also be financially feasible made a project like KCW an easy sell to a state whose residents treasure their natural surroundings and rely on industries that are particularly vulnerable to climate change, such as ski resorts and maple syrup production.

Our group spread out across the turbines' bases like tiny ants amid giant lawn ornaments. In near unison, we aimed our cameras to capture the white whirls against the sky. I surmise we were all performing our own quiet deliberations about the merits of a project that thirteen fellow Vermonters had gone to jail trying to prevent. With the day's light winds and panoramic mountain views, it was not abundantly clear how one or a few sacrificial ridgelines could attract such opposition in a world of oil spills, refinery fires, methane leaks, mining disasters, and rising seas. But my thoughts went back to the question-and-answer session.

The answers to our questions wove a simple and convincing narrative around wind energy in Northern New England. Energy sources have impacts on their surroundings. We saw how the turbines sit on hard-packed flattened areas that were once part of a steep forested ridgeline crawling with wildlife. At night the turbines form a row of blinking red lights that pierce pure darkness in a dizzying rhythm, more like a Montreal dance club than a Vermont nocturne. And although they whooshed delicately on the day we toured, I suspect that most of us, when confronting the turbines up close, could imagine their potential to disrupt close neighbors, as many local newspapers have reported. But in accepting these impacts, neighbors and the electricity consumers of Vermont were opening the door to an era of cleaner electricity production that would benefit society at large. Or so the story goes.

One member of the tour, a short-haired woman in her fifties who drove a minivan and lived in one of the state's more populous counties, asked our tour guide a few provocative questions. She first identified herself as a renewable energy supporter, but she noted the lack of electricity storage options coordinated with the turbine array and inquired pointedly why none had been considered. Before our tour guide could answer, the woman posed a second question. Noting the significant deforestation and the electricity transmission infrastructure already at the site, she asked why every inch of cleared ground had not been outfitted with solar panels, a technology growing cheaper by the day. Her questions were met with somewhat quizzical looks and short, stammering replies. The tour guide indicated that solar arrays were simply not part of the site plans for industrial-scale wind turbines, as if such installations were a standardized cookie-cutter kit that some wind deity implanted on ridgelines. Batteries, he suggested, were not yet commercially viable at such a scale, citing problems at similar facilities that had attempted to utilize them. But he mentioned nothing about adding batteries at a later date should they become viable.

I was intrigued by the woman's insightful questions and by our tour guide's flat responses. What struck me most was their not-so-subtle difference in logic. The questions seemed practical and thrifty, like the penny-pinching Vermont archetype reproduced over generations in my own family, eager to conserve, repurpose, and curtail waste. The tour guide's answers seemed dismissive and couched in the rubric of business accounting, stringing together vagaries and jargon in an effort to deflect from his limited knowledge of power engineering and perhaps from something else. The woman had posed questions using the logic of carbon accounting, and the tour guide had answered using the logic of financial accounting. This interaction confounded me because I had assumed that wind projects were evidence that these two logics were beginning to converge.

Wind proponents maintain that ridgeline wind turbines reduce fossil fuel use, thus limiting carbon emission and mitigating global climate change. Is it not true that producing more renewable power on-site would only further those achievements and perhaps soothe the sting of deforestation? Why would rural electricity projects *not* maximize their productive capacity, particularly when forests have already been lost to the chainsaw? In other parts of the state, residents were lamenting the placement of solar arrays in the state's most productive farm fields and scenic hillsides; so why not integrate the solar arrays to already established industrial locales? It seemed that there was more to the financial and ecological accounting being used to justify ridgeline wind power in Northern New England than what the weavers of wind energy's public narrative were letting on.

In the following months and years, I would learn a great deal from those who had opposed the project most vehemently, including those who ascended the mountain on foot, camped in the cold wet winter months, and went to jail. They believe that the trade-off narrative advanced by wind proponents is far from reality, that it is fabricated and perpetuated in order to advance development projects that show far less promise for addressing climate change than tackling home heating and transportation would. Although I began my research under the assumption that understanding wind opposition was key to reducing it, my focus slowly changed. A more important set of research questions grew from a clearer understanding of Northern New England's particular renewable energy regime. My focus shifted to understanding why and how wind technology generates narratives that mask its complexity and, in so doing, help chart a painfully slow energy transition. So what began as my study to understand resistance to wind morphed into an attempt to simultaneously understand support for wind as well. Instead of focusing solely on why a liberal, eco-conscious region is resisting renewable energy, I ask why this liberal eco-conscious region chose this particular renewable energy path in the first place.

In the chapters that follow, I engage with these questions using lenses from the fields of sociology and environmental sociology. Thus, I devote attention to the regional energy context in Northern New England and the policy environment in which the region's wind projects are sprouting. I illustrate what strike me as the most salient frames for encapsulating the tensions building around renewable energy in North America. This book also strives for a larger goal, which is using sociology as a means of decoding and contextualizing the energy system and energy transition at the heart of all this unrest. While sociological lenses have a way of complicating things that seem simple on the surface, I hope that leaning in to the complexity helps readers at least find guideposts for forming their own opinions about energy and climate change.

Background and Scope

Northern New England encompasses the predominantly rural states of Maine, New Hampshire, and Vermont. It is a region rich with indigenous and colonial history, natural resources, and recreational amenities. Since European settlement, the region's seaside, mountains, and river towns have seen booms and busts in agriculture and extractive industries, hosted enclaves of affluent visitors, and welcomed waves of newcomers and seasonal residents. Within the region's resulting mix of people and institutions are competing development interests hoping to profit from Northern New England's attractive, verdant landscape. The placement of wind turbines across the region's thickly forested ridgelines represents a new but familiar change to the land.

Central to my look at wind energy in Northern New England is the premise that the environmental sociology of renewable energy deserves more critical engagement with the political, economic, and cultural structures propping up the technological apparatus that we understand generally as "renewable." While a great deal of past and current research focuses on energy, this critical perspective has not been prominent within that research. In what follows I will briefly outline the academic literature within which my research is situated.

How Green Is Green?

Social scientists have long studied the exercise and accumulation of political and economic power in the electricity industry. Scholars have documented extensively the structures and practices that investor-owned utilities use to defend their century-old monopoly interests and autonomy.[1] However, surprisingly little social scientific research has endeavored to untangle the web of actors, institutions, stakeholders, and consequences associated with energy systems in the United States.[2] Most studies examine one particular energy type or location of production in depth, but overall we have failed to acknowledge—much less document—enormous variability across American electricity systems and landscapes. Of particular importance to modern environmental issues, social scientists have not engaged the public in contextualizing the structures of corporate and political power from which society's current renewable energy decisions are emerging and have not considered the multipronged implications for global equity and justice. Scholars frame equity and justice very differently across literatures, making some energy "justice" movements seem far less equitable in a global sense than others.

A small but expanding body of literature critically examines contemporary society's transition away from fossil fuels. Some research, particularly that which frames electricity generation as a socio-technical system, has cataloged ways in which renewable energy infrastructure fits into existing electricity regimes structured around fossil fuel extraction.[3] Scholars have observed that wind and solar technologies have been vulnerable to co-optation and scaling-up by legacy energy providers endeavoring to carve new profits from renewable energy markets. A small but growing group of researchers have begun to more critically probe the extent to which renewable energy in its current forms impacts the environment, social justice, and climate change mitigation.[4] This research considers the impacts of renewable infrastructure's manufacturing and eventual decommissioning, in terms of greenhouse gas release, resource use, and social equity. These research threads outline areas where greater forethought is needed to maximize the promise of renewables.

Recent scholarship from environmental sociologists has begun to more critically and more explicitly interrogate the phrase "energy transition," observing that current practices are facilitating energy additions rather than substitutions for fossil fuels.[5] In light of these realities, some energy scholars have even begun to call for a wholesale jettisoning of the term "renewable energy," observing that in practice, the rhetoric surrounding the concept of renewability deviates from meaningful material climate achievements around the world.[6] As climate change worsens, more research is asking why energy transitions underway seem to move so slowly.

Social scientists, most of them working in Europe, have focused heavily on the transformative potential of democratic participation in energy decision-making. Scholars emphasizing "energy democracy" and similar concepts devote attention to the value of energy efficiency that may come from more transparent electricity markets and more participatory electricity planning and governance.[7] For example, smart meters and consumer-engaged tariffs (pricing) and load management have been shown to promote energy conservation and decrease the need for peak generators. In general, this strand of more critical scholarship has not been used to interrogate whether our electricity transition is unfurling with the scale and urgency prescribed by climate change predictions.

To better understand barriers to public participation in energy, we must examine the close entanglement of society's transition discourse and fossil fuel interests. Some scholars still utilize the term "energy transition" with scant acknowledgement that the concept is both a discourse and a business strategy now widely adopted by the fossil fuel industry. "Energy transition" literature tends to highlight social factors facilitating and obstructing moves away from the energy status quo. There is little scholarship probing the energy status quo as fluid cultural phenomenon—what has changed and what has stayed the same in the public's energy beliefs and practices. Transition is something that the public can see reflected in conspicuous landscape change, and many of us assume those changes indicate a shift away from fossil fuels when in fact the energy industry is pursuing additive hybrid models of electricity generation, dependent on fossil fuels. Social scientists must continue to deconstruct, qualify, and specify different transition concepts and narratives, and seek to understand how the public, private, and commercial spheres use and understand those narratives.

Wind, Social Science, and NIMBYism

Northern New England's wind opponents, to whom I will refer interchangeably as wind resisters, have struggled to educate the public about the full spectrum of their concerns because media coverage implicitly promotes the trade-off narrative described earlier: wind energy is far less damaging to the earth than drilling, fracking, refining, or burning petroleum, so it is selfish to reject wind energy because of a few negative sensory consequences felt locally. At the societal scale, it is implied, we should continue to embrace wind energy just as European nations have done for decades. The "local sacrifice for the global good" message presents antiwind sentiment as a NIMBY (not in my backyard) phenomenon, and I propose that approaching wind opponents as antienvironmental from the start also distracts many social scientists from the richness and complexity emanating from antiwind sentiment. Not only should social scientists listen more carefully, perhaps we should also turn our lenses inward. We need to question whether our understanding of wind energy is as superficial as the media's portrayal of NIMBYism.

I have observed over the past five years that many social scientists' approach to studying wind energy suffers from two shortcomings: it treats wind turbine technology as mechanically and economically uniform and as intrinsically beneficial ecologically. Social scientists often discuss wind turbines as if they were simple structures deployed uniformly in multiple sizes, like ubiquitous playground equipment. Wind turbines are typically characterized as limited in their potential only by the wind's variability. However, belying their common mechanics, wind turbines are diverse in how they impact the environment. It would be more apt to see them as complex machinery, like automobiles, than as simple contraptions like swing sets. In addition to being different shapes and sizes, wind turbines serve multiple purposes, utilize numerous backup fuels, and have diverse ownership arrangements and financing models. American wind policies, which incentivize grid-scale wind turbines regardless of their capacity to displace fossil fuels, also adopt the oversimplified view of the technology. In summary, there are many scenarios under which electricity is generated from wind, but only a narrow understanding of those scenarios becomes codified in policy and presented in social science literature.

The tendency to oversimplify wind electricity gives way to a tendency to oversimplify wind opposition; I have observed that this tendency is a pervasive bias among social scientists. Despite great diversity in wind turbines' attributes and uses, applied researchers typically use the "social acceptance" frame, premised implicitly on wind turbines' uniform good. This frame treats wind turbines as fundamentally important infrastructure, compelling most research on wind opposition to simply ask *why* people resist them. It is contrarian to question whether wind turbines may be less good for the environment than we think. Despite the interdependent and intertwined nature of fossil fuel interests in the American electricity landscape, many if not most studies of wind opposition ignore the possibility that opposition is a symptom of a larger problematic energy system, rather than just a problem on its own. Like so many scholars curious about resistance to renewable energy, I too approached my research question first from the "social acceptance" frame. The apparent dissonance between Northern New England's green reputation and local activists' distaste for wind turbines presented a logical puzzle. How can self-proclaimed environmentalists stand in the way of a technology engineered to help the environment?

Countless turbine studies premised on social acceptability yield simple recommendations about distance and money that go unheeded anyway. To summarize, social acceptability studies tell us not to build wind turbines near people's homes and, if doing so is unavoidable, to compensate homeowners fairly and equitably. Several previous studies find that the least contested wind energy projects are those that are municipally owned or that yield widespread community benefits.

While it is important research, the problem with this practical type of inquiry dominating the literature is that it reifies the NIMBY framing—that proximity and self-interest are the primary motivators of dissent—and drowns out other potential approaches. For example, trying to ascertain the price or distance at which wind turbines are tolerable applies a research lens in which turbine critics are concerned with aesthetics or feel entitled to compensation. The same is true for studies about turbine ownership, which implicitly frame criticism as financial. Similarly, as evidence emerges that some home values have indeed been reduced due to turbines, studies might implicate selfishness even when arguing that wind neighbors should be compensated. Without necessarily intending it, operationalizing wind opposition as discrete, measurable variables privileges quantitative dimensions of energy beliefs at the expense of beliefs that are less easily quantified, oversimplifying the global resistance to wind energy as a matter of neighborhood angst.

Wind opposition in Northern New England is disparate and cacophonous, and so I am careful not to interpret it as a cohesive movement pushing
in any single direction. Most social scientists studying contemporary wind
energy in Europe have explained the spread of wind power across Europe as
the result of coordinated grassroots organizing by a pro-environment social
movement.[8] By contrast, recent work in the United States has used the social
movement's fame to make sense of the backlash to renewable energy, energy
technologies, or both. Northern New England's wind industry and subsequent wind opposition are different. The face of the region's antiwind movement is a hodgepodge of organizations and social media pages oriented in
opposition to a single technology, acting as beacons for individuals looking
for affirmation of their distaste for that technology. Still, the region's wind
opposition is a cache of insights worth studying, because in its complexity lie
important critiques about energy systems writ large.

I will draw from several lenses and theoretical schools used in environmental sociology to offer geographic, historical, and conceptual context to
the region's wind resistance. I am not an arbiter of contested and emotional
debates about wind energy's safety or its aesthetic traits. While perspectives
from environmental sociology help to legitimize many criticisms of wind,
my intent is not to pit progressive investors and policy makers against wind
"victims," but rather to frame the region's wind development in a particular domain of environmental capitalism in which ecological accounting is
secondary to financial accounting. To be clear, my position is not that wind
turbines are good or bad. Rather, I propose that absent transparent ecological accounting, the benefits of wind turbines are likely to vary across diverse
topographies, electricity markets, and political economic arrangements. My
research implicates society's energy culture in obfuscating turbines' ecological impacts and highlights the need for greater access to and openness in
energy conversations so that the renewable transition unfolds in sync with
society's climate imperative.

In summary, antiwind research should methodically correct for pro-wind
bias, and even more broadly, this research should take stock of its original
research questions, which can inadvertently reproduce bias unbeknownst
to the researcher. We assume that wind turbines produce energy in a way that
minimally impacts the landscape, which affords us personal affirmation
and a modicum of environmental optimism. For the record, I consider
myself no different. In the course of conducting the research that went into

this book, I have had to continually assess how what I read, hear, and observe resonates with my operating assumptions: foremost, the assumptions that wind turbines are always good and that opposing them is always bad and thus deserving of study.

This book is not about what motivates antiwind activists or how they might successfully unify or better mobilize support for their fight. As I said, Northern New England's wind resistance does not fit the profile of a social movement, but I wish that it did. Before a real social movement can build around energy transition, there needs to be more work exposing the terrain where high-impact productive change can get a foothold.

Instead, I show how wind resistance offers an entry point for explaining how and why the country's energy transition remains captive to orthodox decision structures and is thus proceeding woefully slow. One of this book's central arguments is that energy systems' technological complexities juxtaposed with wind turbines' aura of simple purity explain why climate-aware citizens are predisposed to support wind. Unpacking wind energy's political, social, and economic context, helps expose the fallacy of presuming that such a widely and complexly executed technology is uniform in its impacts. Industrial-scale wind developments nest within an opaque miasma of distant parent companies and shareholders, many of whom have interests in fossil fuels. I advocate for new approaches to thinking about this status quo energy system. To that end, I continue to widen the concept of "energy transitions" to include alternatives to that which is outlined by global energy hegemons.

Methods

I tried to follow a simple overarching research question in writing this book: Why does Northern New England's wind development meet with social and political resistance, despite the region's environmental progressivism? However, the complexity in Northern New England's renewable energy discourse suggested that a simple research question would not beget a simple answer. In endeavoring to provide an answer that is useful and transparent to a broad audience, it has been my intention to disentangle the origins of and audiences for opposing narratives in the region's debate over renewable energy.

My data come from several sources. To begin, I conducted twenty in-depth semistructured interviews with key informants in the fields of business, politics, and activism and advocacy. The interviews began with a series of questions relating to the informants' history in the region, their experience with ridgeline wind turbines, and their experience with energy more broadly. I wanted to contextualize my informants' perspectives on wind energy within their experience of the region and their connections to its landscapes and communities. These interviews ranged from ninety minutes to more than four hours in length, and some involved short visits by car to sites of relevance to local energy. My interviews served two purposes. Foremost, the interviews oriented me to central structures, actors, and institutions involved in the issue of renewable energy in the region; the interviews also helped corroborate and elucidate dominant themes in the public conversations that transpire around renewable energy. Since conducting my first interviews, I have followed up a few times with some of my informants to reflect on my findings. Despite a willingness by most interviewees to talk "on the record," I have not disclosed their identities or attributes that would make them identifiable.[9]

Building from my first interviews, I have observed the public renewable energy discourse through four channels. First, to observe discourse embodied in interpersonal exchange unfolding in real time, I attended approximately fifteen public renewable energy forums and events in all three states in person and viewed publicly available video recordings of another twenty of these events. At the events that I attended in person, I both used an audio recorder and kept field notes that I soon typed into longer thematic reflections. Transcripts were unavailable for many of the publicly recorded events, requiring me to take field notes as I viewed them, as I would have done if I had attended them in person.

Most of the events I have attended and observed are highly structured, in that they were moderated by a figure of some authority and promoted a specific type of testimony. Public events put on display a particular back-and-forth dialogue between policy supporters and opponents. Those who wish to speak must first register themselves on a sign-in sheet; they are then called to the front by a moderator in the order in which they signed in. Without local insight, the nature of each speaker's testimony is unclear until the first few words are uttered. One either endorses or mounts a case against the wind project under discussion. Though rural New England is known for its tradition of town meetings, the formats used for eliciting energy feedback seem foreign. The tone conveys a sense of urgency and the procedure dispenses with any idyllic democratic imagery. Testimony is recorded, sometimes by court stenographer, to create a record for decision-making removed from public sight and occurring at a later date in the state capitol.

Some of the events I observed were formal stakeholder-engagement meetings, which typically satisfied an entity's minimum statutory obligations to involve the public in the process of siting new energy installations or in considering policy changes related to energy. Other events that I attended superseded the minimum obligation. These include the public meetings sponsored by developers to build support for their proposed projects. I also observed, for example, expert and public testimony before legislative committees, as well as testimony before public boards.

As others have noted, the tenor of media coverage has changed over the course of the past several years, and public events have changed accordingly. With more wind projects under development and more media outlets covering the political controversy surrounding them, turnout at public meetings seemed to grow. Tension seems to have ratcheted upward. More recent events struck me as public airings of grievances, primarily by wind opponents, bitterly aware that their testimony does not constitute a vote in a majority-rules scenario. Testimony sounds more like strategic posturing that resonates with common talking points and seems as though it is leveled to elicit support from others in the audience. Representatives from the renewables industry appear hardened and ready for criticism. Their ranks now include public relations professionals, consultants, executives, and lawyers.

Second, over the course of my research between 2013 and 2019, I have cataloged more than one thousand newspaper articles focusing on energy and renewable energy from the region's print media, which I have supplemented with articles dating back to 2000. This sample includes news reports relating to renewable energy and transmission from major regional newspapers and smaller, community papers near wind developments. As media coverage has broadened, a slow shift in coverage has been led by news outlets slowly engaging with alternative frames for the region's renewable energy controversies. This shift has been led by nationally circulated coverage by the *New York Times* and Vermont's statewide independent online investigative news website, the *VTDigger*.

Third, the richest narrative data I have encountered has appeared in virtual space, in two online formats: the *VTDigger*'s comments section, which requires contributors to register with their names, and public and private Facebook pages maintained by regional nonprofit organizations and more loosely assembled groups of activists. Whereas the Facebook pages provide a forum for movement-building and the broadcasting of information salient to local and regional wind projects, the *VTDigger*'s comments section fosters dialogue among readers with multiple perspectives. The Facebook pages are akin to an online pep rally, and the *VTDigger* data offer a veritable focus group, wherein more diverse commenters respond to provocative reporting and to one another. For most of the *VTDigger*'s existence, its comments section enforced no limit on the quantity of text accepted from each contributor, which means that many reader comments are thoughtful and in-depth. In fact, dialog between commenters sometimes references academic research and links to data that support their positions. A few of my interview informants are also regular contributors to the forums and Facebook pages, though generally these virtual communities include many others with an interest in regional energy. Facebook groups include both individuals living in the region and those with a direct interest in the region who live elsewhere, primarily owners of seasonal homes. Bearing in mind how sparsely settled most turbine-hosting communities are, these forums are probably quite representative of the communities' spectrum of wind supporters and opponents. Conspicuously missing are people in the middle.[10]

Lastly, I observed the region's renewable energy rancor through eyes and ears that are well acquainted with the area, but fresh from several years spent away. I was born and raised in Vermont and have family roots in all three northern New England states stretching back to the eighteenth century. After finishing college in Maine, I left the region for eleven years but returned for four years in 2012. My upbringing makes me intimately familiar with regional culture and political geography, and so I brought to my fieldwork the intrigue, investment, and entrée of an insider.

Living and traveling extensively throughout the region over the course of four years immersed me in a visceral sense in debates over renewable energy. I have, of course, visited wind projects up close and documented the discourse emanating from various public spaces. But without any effort, the issue pervaded my "private" life as well. I own property in Vermont and have had innumerable community interactions there that serve as informal data points. The community in which I was raised, however, is in a county without existing or proposed turbine development, and my friends and family who remain in the area do not have intimate knowledge or opinions about renewable energy development derived from anywhere but statewide media or word of mouth. Describing my research to friends, acquaintances, and complete strangers elicited countless discussions and commentaries, several of which I would document before I reached saturation; I realized quickly that casual observers of the region's energy issues tended to have opinions of wind projects that fall along a narrow spectrum of zeal, indifference leaning toward support, or ire.

Traversing the region by car, I observed new projects under construction conspicuously perched atop ridgelines and tuned in inadvertently to radio call-in shows debating the issue. I observed renewable energy discourse unfolding slowly in the form of yard signs and installations displayed to combat, but sometimes support energy infrastructure in the region. I have documented these in photographs and included some throughout the book.

My perspective is also cultivated through extensive experience abroad, through my training as a sociologist, and through living and working in rural regions across the upper Midwest and the Pacific Northwest. I reflect here on this status as both insider and outsider transparently, so that readers can assess my findings for themselves with full knowledge of my personal and academic origins. Put simply, I believe that my deep familiarity with the region's demography, history, politics, economy, culture, and identity is an asset in presenting Northern New England's renewable energy controversies to a broader audience.

The Chapters Ahead

Chapter 2, "Windy Ridgelines, Social Fault Lines," explores how wind projects agitate animosity. The chapter draws contrasts between three types of arguments: those that are superficial and based on aesthetics; those that are based on technical aspects of turbine finance, carbon accounting, and ecological disruption; and those that are based on procedural exclusion from energy planning and regulation. Chapter 2 encourages readers to look for and sift through complex arguments that might otherwise be obscured by labeling wind opposition "NIMBYism."

Chapter 3, "For the Love of Mountains: The Green Politics of Place," explains the historical conditions and policy developments that gave rise to Northern New England's ridgeline wind turbines. The chapter shows how the region's pleasing landscape became a crossroads for tourists, second-home owners, and urban newcomers desiring to build their lives against an unspoiled backdrop, making many sensitive to outside influences. Chapter 3 describes how public discussion of renewable electricity has been scripted by the particular way the region's land, people, and history interact, offering lessons for emerging rural-urban energy tensions more broadly.

Chapter 4, "But What If . . . ? Wind and the Discourse of Risk," helps explain why finding solid data about wind energy is so difficult. The chapter explores confusion and uncertainty as tactics that both advance and oppose ridgeline wind turbines, illustrating German sociologist Ulrich Beck's theory of the "risk society." Risk society entails increasing societal orientation around fear and uncertainty about the future and it can be seen in the skepticism of expertise and authority in the wind debate. The chapter focuses on electricity pricing and carbon accounting to illustrate how actors mobilize discourses of risk and use conflicting scientific evidence to frighten the public about the uncertainties behind energy decisions.

Chapter 5, "Following Power Lines: A Political Economy of Renewables," unpacks the social structures and relationships aligning to make Northern New England's wind turbines profitable. The chapter contains two parts. The first focuses on money—how new policies following neoliberal patterns helped institutions exploit turbines' financial architecture to the institutions' advantage, with little emphasis on the environment. The second part focuses on people—situating the region's wind industry amid a coalition of political and private interests acting in concert to fulfill professional obligations outlined by antiquated, growth-oriented energy policy. Chapter 5 helps contextualize wind opponents' accusations of corruption and feelings of exclusion from energy planning.

Chapter 6, "Scripted in Chaos," explores how wind turbines fit into a protracted national energy transition designed with the needs of fossil fuel interests in mind. These interests, with large investments in natural gas and navigating chaotic trends in the energy sector, benefit from the wind's intermittency because they have positioned natural gas as a clean alternative to coal and oil that can be dispatched quickly when the wind ceases to blow. The chapter describes how natural gas became nationally celebrated as the ideal "bridge fuel" despite the disastrous impacts of fracking and mounting uncertainty about gas's climate consequences. The chapter then explores how the resultant road map to a slow transition gains traction from rhetoric demanding justice for oil and gas workers while ignoring those workers' relatively privileged status in the hierarchy of fossil fuel victims.

Chapter 7, "Why We Follow the Slow Transition Road Map," examines how fossil fuel interests elicit tacit support from the public for their vision of a slow transition. The chapter considers (1) the origins of energy culture in industry propaganda and the creation of docile consumers, (2) the supporting role that wind turbines play as totemic icons of environmentalism obviating critical engagement in energy issues, and (3) the inability of promises of green jobs to fill the important role of sustaining rural communities. This chapter contends that these conditions weaken and distract from meaningful energy engagement and thus deter us from pursuing superior energy alternatives.

Chapter 8, "Ecological Modernizations or Capitalist Treadmills?," uses two opposing theories to help readers see ridgeline wind's simultaneous strengths and weaknesses. The chapter situates Northern New England's experience in ecological modernization theory (EMT) and the treadmill of production (ToP), oppositional theories in environmental sociology. The chapter describes how ridgeline wind can be understood optimistically as environmental progress, and at the same time, pessimistically as a business-as-usual Band-Aid. Chapter 8 considers evidence from Vermont, Northern New England's most environmentally progressive state, to show the complexity belying popular public narratives of eco-progressivism.

Chapter 9, "Energy and 'Justice' in the Mountains," exposes how global injustices are complexly imprinted in first-world disagreements about renewable energy. The chapter argues that Northern New England's long history of white environmentalism has endowed it with abundant commitment to ecological causes but little capacity to build bridges to social causes. Chapter 9 focuses on how participants in the wind debate invoke "justice" to support their case by appropriating others' injustices, but not meaningfully addressing or remedying those injustices.

Chapter 10, "Reimagining Energy," reflects on Northern New England's wind debate to outline strategies for more just energy planning. The chapter argues that prioritizing any energy for its low costs can exacerbate persistent injustices wrought by overconsumption and widening inequality. The chapter emphasizes that transparency and greater citizen involvement in energy planning are necessary for advancing energy solutions outside the status quo. Chapter 10 works toward an understanding of how alliances across communities and issues might promote inclusivity while advancing opportunities at smaller, more efficient scales.

Final Warning

Though readers may be looking for a definitive stance on wind turbines, the social sciences are better at exposing and characterizing complexity than making it sit well with us. While it sheds light on the wind turbine debate, this book cannot offer the quick and uncomplicated answer that most of us hope for when making a decision about a complex issue. I try to tell the story of Northern New England's wind turbines by highlighting a few main characters and a dominant narrative arc with several subplots and minor scenes, not simply two opposing sides. Both wind opponents and wind developers are diverse groups, as are policy makers. It would be impossible to pinpoint exactly who or what is authoring Northern New England's energy story or how it will end. Nonetheless, sociology is good at offering ways to think about real-life cases—ways of thinking that may help individuals arrive at the conclusions they seek or perhaps arrive at contentment with the complicated reality that they see through a new light.

Erecting wind turbines on mountains can be both a carbon-friendly first step in the long term *and* a carbon-intensive boondoggle in the short term. Similarly, the driving forces behind turbine construction can be both ecologically intentioned and profit-seeking. Rather than claim that one perspective is more accurate or truthful than another, this book proceeds by showing how wind turbines fit into a particular place and time. Again, the book's motivating and overarching objective is to explore the apparent contradiction between strong environmentalism and opposition to wind energy. Put simply, why is such an environmentally progressive region reacting negatively to a technology associated with protecting the environment? Each of the following chapters answers this question with increasing complexity, culminating in a recommendation for greater public consciousness of and involvement in energy.

2

Windy Ridgelines, Social
Fault Lines
● ● ● ● ● ● ● ● ● ● ● ● ● ● ● ● ● ● ● ●

A Lunch Room United, a State Divided

In autumn 2016, I attended a public comment session for Vermont's new energy plan in a middle school cafeteria just outside of Burlington, the state's largest city. To most Americans, a Vermont suburb would seem like a small town. Essex Junction is, however, a small town fortunate enough to offer residents high-paying jobs at a semiconductor factory that IBM built there in 1981. Vermont's new energy plan updated a 2005 policy with longer-range renewable electricity targets and greater attention to carbon reduction. Public comment sessions were held throughout the state to solicit feedback from Vermonters.

As I had learned to expect, the parking lot was a sea of Subarus, Volvos, and hybrids. Along the pathway from the parking lot to the school entrance, antiwind demonstrators had planted four or five small signs in the ground. On the other side of the path, along the school's brick exterior, a row of demonstrators dressed in black stood silently, holding handmade posters, with black coverings draped over their faces to symbolize feeling voiceless.

I sat with about fifty others in plastic chairs to learn about the new energy plan and to hear what Chittenden County residents think about

FIG 2.1 Wind protesters stand silently against a wall at Essex Middle School, Chittenden County, Vermont, outside a public meeting on the state's revised energy policy, October 2015 (Photo courtesy of Shaun A. Golding)

it. I expected the county's more affluent and "urban" residents to be aggressively progressive, and they were. Several commenters lobbied passionately for a statewide carbon tax, a far more progressive policy recommendation than I had heard anywhere else in the region, and indeed more progressive than what any other U.S. state has been able to legislate.

In the more rural parts of Northern New England, I had grown accustomed to anger over wind turbines and to exasperation with lawmakers and regulators. Despite walking past the wind opponents standing in quiet defiance outside, the evening's speakers paid scant attention to the state's ongoing antiwind activity. One speaker expressed hope that wind conflicts could be resolved, but nearly all others were there to celebrate the state's energy planning achievements and to push for more aggressive efforts. It seemed to be a foregone conclusion that wind opponents, including those lined up just outside the door, were merely a small faction with extreme, impractical, and emotional views.

The evening's testimony made the prospects for progressive change seem bright. That was of course before November 2016, when Donald Trump would win the presidency, and Vermont's pro-wind Democratic governor would be replaced with an antiwind Republican. Wind development would eventually slow to a halt. Although lawmakers' environmental commitment has in many ways been amplified since the 2016 election, it has been kept in check by the Governor, who has twice won reelection.

What seems most clear in hindsight is that Vermonters in the state's more affluent urban enclaves may have missed something by failing to engage with their more rural antiwind neighbors. While the region is progressive and environmentally conscious, it is home to many shades of environmentalism.

Contentious issues have many sides and give rise to many stories, some of which go unheard but find their way out somehow. I want this book to help make sense of the vexing question, Why are environmentalists responding negatively to renewable energy? In this chapter I start to answer that question by explaining why people feel silenced in the wind debate. I start to make the case, which continues throughout the book, that we need to hear more voices in order to understand why wind energy is unpopular. I would also like this book to show how sociology, and particularly environmental sociology, might help readers envision their personal engagement with energy issues in a way that reaches below the surface of energy debates, is self-reflexive, and respects difference and diversity.

Northern New England and Renewable Energy: The Basics

It is only recently that politics, economics, and technology aligned to make wind energy viable in Northern New England. Like most places in the United States, the region's major electric companies began as investor-owned monopolies that owned dams, power plants, and power lines to deliver electrons to customers' homes. Without petroleum reserves and having already dammed many of its rivers, the region looked to nuclear power in the 1960s and 1970s. Each state oversaw the development of one nuclear facility to meet much of their own electricity demand, plus a surplus for export. While nuclear energy was of course controversial, the states located their facilities along their southern boundaries, away from their population centers, to allow easy delivery to cities and to take advantage of large water bodies for cooling. Now, with two of the region's three nuclear facilities decommissioned, the region must import electricity.

Despite most voters' preference for renewable energy,[1] Northern New England has few options to make renewable energy profitable for the region's legacy electricity interests. Northern New England is not especially sunny, and its mountains are not easily built upon. Maine, a much larger and sparsely settled state by area, is better positioned for wind than Vermont and New Hampshire. Maine ranks sixth among U.S. states for installed wind capacity and derives nearly a third of its electricity from in-state hydroelectric power. But throughout the region, dams are old, and in some rivers, old dams are being removed to restore decimated fish stocks. Having been denuded of nearly all its timber, farmed intensively, and then reforested in the course of 150 years, the three-state region is now somewhere between 75 and 85 percent forested. Its biomass resources constitute a sizable portion of its energy portfolio both in home heating and electricity generation, but wood-burning power plants typically rely on waste from the lumber industry, so they are not primed for rapid expansion; nor are they ideal for reducing greenhouse gas emissions.

The region is adjacent to the Canadian province of Quebec, where several enormous hydroelectric projects have offered Northern New England electricity customers supplemental power for decades. However, only Vermont buys significant quantities of power from Quebec, and the deficit left by closure of the region's nuclear reactors is now met primarily by gas power plants, which are fueled by the abundant and cheap fracked natural gas piped from the Marcellus Shale in Pennsylvania, West Virginia, and Ohio. During peak demand on the coldest winter days and the hottest summer days, the electricity grid also utilizes foreign gas shipped through Boston by boat, or oil, to run the region's space heaters and air conditioners. Two coal power plants in New Hampshire are available under the heaviest demand scenarios.

At present, wind and solar are widely regarded as the preferred renewable electricity options, because flooding forests and burning wood biomass emit greenhouse gasses and are forgone opportunities for sequestering carbon dioxide. With an eye toward advancing wind and solar power within the three states, leaders and voters have nurtured a modest renewable energy economy since the 1990s. Legislators in all three states moved to ensure that they could benefit from federal subsidies for wind and solar projects.[2] By the time those subsides peaked under the 2009 American Reinvestment and Recovery Act, also known as President Obama's stimulus plan, the region had one successful ridgeline wind farm in operation for more than ten years and several others planned.

Northern New England Electricity Comparison			
	Maine	New Hampshire	Vermont
Population/Land Area	1.34 Mill./ 30,862 sq mi	1.36 Mill./ 8,972 sq mi	624,000/ 9,250 sq mi
Electricity Market	Deregulated	Deregulated	Regulated
% Served by IOU	90%	87%	80%
Rank Electricity Use per Capita	41 of 51	44 of 51	42 of 51
Net Electricity	Importer	Exporter	Importer
Primary Renewables (% of all renewables)[1]	Wood/Waste (29.4%) Hydro[2] (46.9%) Wind (23.6%) Solar (.01%)	Wood/Waste (54.6%) Hydro[2] (35.3%) Wind (10.1%) Solar (<.01%)	Wood/Waste (11.1%) Hydro[2] (81.0%) Wind (6.2%) Solar (1.8%)
Rank Electricity CO_2 Intensity	46 of 51	48 of 51	49 of 51
Renewables as % of Generation 2018	84.2%	39.9%	98.1%
Relaxed Wind Siting	Since 2008	No	Since 2005
Grid-tied Wind Sites (turbine count)	18 (360)	5 (83)	5 (78)
Pre-REC Wind MW (% of total MW)[3]	923 (19.8%)	185 (2.3%)	149 (13.4%)
Rank Electricity Price	10 of 51	6 of 51	8 of 51
NE Regional Grid Fuels (% of total)	Nuclear (33.5%) Natural Gas (38.1%) Hydro[2] (12.4%) Wood/Waste (9.0%) Wind (3.3%) Coal/Oil (2.6%) Solar (1.2%)		

[1] 2018, pre-REC trade
[2] Domestic and Canadian hydro combined
[3] 2019
SOURCE: https://www.eia.gov/state/seds-data-complete.php

The states' wind energy policies followed different paths. Maine instituted a renewable portfolio standard (RPS) that set a minimum quantity
of renewable energy for Maine electric companies to source. The Maine
legislature also passed an ordinance to make the process of application and
permitting transparent for municipalities establishing their own energy siting rules. New Hampshire adopted an RPS in 2007, and in 2015 the state

Northern New England Electricity Time Line
1972 - Vermont Yankee Nuclear begins generation
1972 - Maine Yankee Nuclear begins generation
1976 - Seabrook Nuclear commissioned 1976 (bankrupts utility 1988, begins generation 1990)
1987 - Maine enacts net-metering
1987 - Vermont buys first electricity from Hydro Québec
1992 - Federal production tax credit for wind initiated
1996 - Maine Yankee Nuclear ceases operations
1997 - Region's first modern grid-tied wind installation opens in Searsburg, Vermont
1998 - Vermont, New Hampshire enact net-metering
1999 - Maine sets renewable portfolio standard (40% by 2017, 80% by 2030,100% by 2050)
2002 - Quebec government signs agreement with Grand Council of the Crees
2003 - Maine passes Wind Energy Act
2005 - Vermont enacts SPEED law (voluntary renewable targets, promotes wind with RECs)
2007 - New Hampshire sets renewable portfolio standard (25.2% by 2025)
2008 - Maine's unincorporated lands designated expedited wind permitting area
2009 - American Reinvestment and Recovery Act (AKA Obama stimulus) boosts wind
2011 - Vermont ramps up net-metering to 20 cents per kilowatt-hour
2012 - Wind resisters arrested protesting Kingdom Community Wind
2012 - Maine adopts tighter wind turbine noise setback standards
2014 - Connecticut passes law forbidding its utilities from buying Vermont RECs
2014 - Vermont Yankee Nuclear ceases operations
2015 - Vermont Act 56 institutes renewable portfolio standard (55% by 2017, 75% by 2032)
2015 - New Hampshire enacts wind siting policies
2015 - Maine allows for removal from expedited siting by petition
2015 - New Hampshire adopts wind turbine noise setback sound standards
2016 - Maine rolls back net-metering
2016 - Vermont enacts Act 174 for municipal energy planning; tightens turbine noise standards
2016 - Maine governor places moratorium on wind in scenic places
2016 - Massachusetts issues call for 9.45 million MWh of renewable energy
2016 - Maine eliminates net-metering policy
2017 - Vermont reduces net-metering 13% for residential, 35% for large installations
2018 - Maine reverses wind moratorium
2018 - Northern Pass defeated by New Hampshire regulators
2019 - Maine reinstates net-metering, Creates climate council to meet emissions targets
2020 - Vermont lowers net-metering rates, Creates climate council to meet emissions targets
2020 - Maine citizens fight New England Clean Energy Connect corridor project
2020 - New Hampshire debates large net-metering

FIG 2.2 Time line of regional energy events

finalized its approval process for siting wind turbines. Vermont's energy planning took a controversial tack, despite the state's sterling green reputation. Vermont's 2005 Sustainably Priced Energy for Economic Development (SPEED) policy aggressively promoted turbine construction to satisfy other states' RPSs. Accordingly, SPEED proved superficial in its climate impact compared with a typical RPS, which replaced it in 2015. Also, Vermont

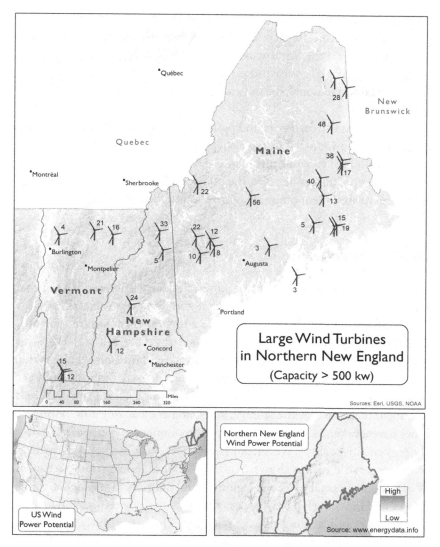

FIG 2.3 Map of existing wind installations (Source: US Geological Survey, energydata.info)

legislators exempted energy developments from the state's strict development review process and instead diverted them to the PUC, formerly known as the Public Service Board (PSB).[3]

A construction boom ensued in the region as technology and federal subsidies converged. The number of new wind permit applications more than doubled in Vermont between 2007 and 2011, climbing from seven to seventeen.[4] Between 2008 and 2018, Vermont saw the completion of four grid-scale wind projects. Developers completed sixteen projects in Maine and have proposed the installation of several others, including one with as many as 119 turbines. Four wind projects came online in New Hampshire. By 2015, Maine had doubled wind energy's representation in the state's electricity portfolio.[5]

I will address the corporate and political architecture of Northern New England's wind industry in later chapters. For now, I will focus on the region's resistance to wind turbines, to lay the groundwork for more substantive observations about how the policies driving the American energy transition relate to economics, politics, climate change, and justice.

Wind Resistance

Environmental sociologists Simon Guy and Elizabeth Shove observed that for a very long time, electricity was just not an "emotive subject" for most Americans.[6] Guy and Shove wrote in 2000 that "energy consumption fails to make the grade as an item of popular concern." For most of my life, this was true in Northern New England. For decades, Maine, New Hampshire, and Vermont put their primarily nuclear electricity nebulously onto the regional grid, with sporadic debate focusing primarily on the perceived threats of meltdown, waste storage, and water release. But now, wind turbines have awakened "emotive" dissent from numerous perspectives across the region. This backlash is evident in newspaper accounts, political rhetoric, road signs, radio call-in shows, YouTube videos, and election outcomes. Wind opponents challenge the regional wind industry in the media, in the courts, and in voting booths. Opponents speak out at meetings and forums at every stage of planning and site selection. Opponents rally against the meteorological towers that gauge wind's potential before a project's financing is even secured or ground is broken. Wind opponents have tried to stop construction in its tracks, and they police the turbines when they are brought online by measuring their sound and documenting their interference with wildlife. This active, popular engagement with energy issues has come suddenly. It is both hard to miss and hard to make sense of in an otherwise progressive environmentally conscious region.

Like Pasta, Arguments That Stick

As I learned in my interviews, the more someone studies wind energy, the more arguments they lodge against it, which they do in rapid-fire bursts. This is the pasta method: opponents try to see what arguments stick in the same way one might throw spaghetti at the wall to see whether the pasta is fully cooked. A few activists in particular have amassed a wealth of energy expertise, so talking to them about ridgeline turbines uncovers several dozen reasons to oppose the turbines. As I will discuss further, nuanced critiques are not always separated or easily noticed when opponents utilize the "pasta throwing" method of public relations, and these critiques certainly are unclear for journalists and the public.

What wind opponents lack in regional cohesion they overcompensate for with a shared global antiwind gospel, which might make wind opposition seem more organized than it is. A worldwide antiwind community has formed online, through LISTSERVs, social media, and websites like windwatch.org, which compile news about wind energy's problems and failures. Constantly bombarded with negative media attention, the wind opponents tapped into these virtual communities have a deep arsenal of arguments against industrial-scale wind turbines. Studies suggest that exposure to antiwind sentiment strengthens one's own sentiment.[7]

Opponents' concerns can be placed in several overlapping categories, outlined in the following paragraphs.

Surface Tensions

Criticisms related to turbine mechanics and proximity are the easiest to grasp and tend to be the only complaints recognized by regulators. Therefore, these criticisms usually rise to the surface of popular media and news reporting.

Many, though not all, domestic and international survey respondents would prefer to live far away from industrial-scale wind turbines.[8] Some research finds that this preference is particularly evident in highly valued landscapes.[9] This finding makes sense. Many modern wind turbines would dwarf the Statue of Liberty. They cast flickering shadows, blink asynchronously at night, throw ice, and make noise of varying magnitudes. Turbines are assailed for being fire hazards, leaking oil and coolant, and killing birds and bats, and for the difficulty of their decommissioning, removal, and disposal.

People around the world live close to industrial-scale wind turbines, and many if not most see no problem with them. Research shows that sharing ownership or receiving compensation for hosting wind turbines is associated with accepting them.[10] These findings are ubiquitous and obvious. Naturally, one feels better about making sacrifices for society when society acknowledges it. And yet industrial-scale wind projects proceed around the globe adjacent to homes and recreational spaces, and antiwind sentiment continues to snowball.

Reading newspaper accounts from Maine, New Hampshire, and Vermont, reveals plenty of turbine neighbors who are dissatisfied with mechanics and proximity. Some close neighbors complain of irritation and sickness resulting from the noise and subtle changes in air pressure that accompany the rhythmic whirling of the turbines.[11] These issues have displaced multiple residents from their homes in Northern New England and have garnered significant media attention as a result.

The world is divided as to whether wind turbines are attractive to the eye, and whether they lower property values has been a contentious offshoot of the aesthetics debate. Local evidence of lowered property values suggests that, like surveyed homeowners, home buyers also prefer a buffer between wind turbines and their homes. This becomes evident when homes located near turbines take longer to sell and go for less money.[12] While many studies find no evidence of lowered home values, these results could be a matter of research design.[13]

I describe these surface complaints first because I want to distinguish them from critiques about how turbines operate as technological systems, interconnecting with political and economic structures. Because of the diverse social and political channels through which wind projects move from test site to construction, understanding wind opposition requires deeper exploration of how opponents perceive the context from which projects emerge—not simply how opponents perceive wind turbines' physical traits.

Below the Surface

Northern New England's debate over renewable energy is noteworthy in part because of who many of the antiwind advocates are and what they believe. Several are lifelong environmentalists and, irrespective of their initial motives, have developed nuanced critiques of ridgeline wind energy. Many are career environmentalists who, before their opposition to wind turbines, held a high-degree of professional credibility. In staking their positions against wind turbines, many antiwind environmentalists find themselves in conflict with pro-wind voices with whom they had long been allied.

From the start of my research, it seemed implausible to me that career environmental professionals would sacrifice their credibility to save the way a mountain looks or because they believed what they read on an antiwind website. So I sought out some of the region's activists to learn more about why they had put their credibility on the line. Most of my interviewees moved to the area between the late 1960s and 2000, and none can see wind turbines from their homes. Some had engaged in direct actions to halt wind projects, attracting media attention that made them easy for me to locate. But as I will explain, their efforts did little to expand the public's understanding of wind opposition beyond surface issues.

In addition to the issues of mechanics and proximity, antiwind messaging deploys arguments that I categorize as technical or procedural. Technical arguments encompass critiques of ridgeline wind's lack of financial and environmental transparency. Procedural arguments relate to the barriers that wind opponents have faced when trying to shape energy policy with their technical arguments. Whereas technical arguments raise issues with the efficacy of energy systems, procedural arguments raise issues with undemocratic planning and regulatory structures that facilitate energy systems. These arguments were shared with me in several long interviews over the course of a few years. As I explain later, wind opponents struggle to effectively broadcast the full breadth of these arguments to the public.

Technical Antiwind Arguments. Northern New England wind opponents advance detailed arguments for why industrial-scale wind is ill-suited for the region's ridges, electricity grid, and carbon reduction. Most of the resisters I interviewed have done their energy research and concluded that ridgeline wind turbines in New England are not transparent in reducing carbon emissions. And while wind energy now presents a partial alternative to fossil fuels, wind opponents believe that large wind farms will not be the backbone of our future electricity system. What I describe in this section will form the basis of subsequent chapters, so for now I outline only a rough sketch of resisters' technical arguments.

Many wind opponents believe that the region's ridgeline turbines do not make climate sense. Citing wind's intermittency and a lack of electricity storage, these resisters observe that each new turbine requires proportionate fossil fuel backup for when the wind is not blowing. One estimate is that 1 kWh of wind and solar requires 1.1 kWh of backup.[14] Wind opponents contend that the true impacts of backup power are closely guarded, raising doubts about the actual carbon displacement each turbine can offer. My informant Karen told me, "Come to find out, nobody's actually paying attention to what the impact is! Apparently they don't care. They're just concerned about the economics or the politics. Promoting renewables sounds so good to everybody. That's what we're doing. We're just asleep at the switch."

To be clear, the wind opponents I've talked to are not disputing that industrial wind turbines displace fossil fuels from the electricity grid. Opponents are calling attention to the fact that industrial wind development proceeds with no real accounting of how much, where, and when those reductions occur, and how they compare in cost and impact to alternative climate change mitigation strategies. In Northern New England, wind opponents argue that as tools for climate change reduction, ridgeline wind turbines are inefficient compared with turbines placed elsewhere and compared with measures targeting the region's car dependence and heating, which account for the wide majority—as much as seventy percent or more—of the region's greenhouse gas emissions.

Wind opponents contend that generous federal subsidies in the form of tax credits for developers, many going to foreign companies, make the region's ridgeline wind turbines artificially feasible. Some opponents have undertaken studies that refute the cost projections used to justify wind projects' cost effectiveness. These opponents articulate a cogent critique of electricity bottlenecks resulting from mismatch between turbine siting and local grid capacity, noting that on days with high electricity demand, turbines are taken offline, spinning idly because the grid cannot accept their power or because grid operators prefer less volatile sources under high-demand conditions.

Further, opponents contend that remote ridgelines are ecologically costly sites on which to build because of the blasting, flattening, and deforestation required. I discussed this argument with my informants Ralph and Jim one gray summer day. We sat in a town hall under natural light flooding our table from an enormous window. Ralph told me that it was at that very table that he first learned of the wind industry's entry to the region. Residents from a community targeted for wind development attended a meeting to warn citizens in Ralph's town, which had also been targeted. It turned out the residents of the neighboring town were primarily warning about aesthetics, but Ralph did not care about aesthetics. Only after he volunteered to read the developers' proposal for his town did he realize that there would be impacts to ridgeline ecological systems, too. Buried in the 1,300 pages were engineering diagrams showing filled-in streams and forty-foot-deep cuts to the ridgeline bedrock.

The project proceeded despite vehement local opposition in neighboring towns. Ralph scrolled through some high-resolution before-and-after photos and said to me, "I was entertaining fantasies of helicopters flying the turbines in and putting them down in the woods and cutting a few trees around them, and then taking them out in 20 years, not blowing up the ridgeline." As a result of these construction practices and their outcome, wind opponents express grave concern over storm water runoff, because before the blasting the lush ridgeline was an important buffer to heavy rains.

The ridgeline in Ralph's town was the last undisturbed corridor for wildlife for a few hundred miles, which made the loss of habitat a pressing concern, too. Given the significance of that specific location, local opponents impugned the ecological logic of land swaps that traded the developed land for noncontiguous parcels that offered little benefit to the displaced wildlife. Since the project's construction, resisters have also assailed state rulings that relaxed protections for a threatened bat species falling prey to the turbines' spinning blades.

Opponents also question renewable energy's new geography, which positions rural communities as the sites of power generation for more urban places. In Vermont in particular, opponents deride the state's energy laws for having a looser definition of renewability than any other state in the region and for allowing Renewable Energy Certificates (RECs) to be sold to other states. RECs certify each electron that is generated by renewable energy as special. Progressive states and cities now mandate that electric utilities include these special electrons in what they sell their customers, which created a market for RECS to be traded among electric companies across the country. (As I will detail in subsequent chapters, RECs are the shaky ground on which the United States has erected most of its climate strategies in the electricity sector.) In Vermont's first pass at a renewable energy law, an in-state utility could sell all of its RECs to an out-of-state utility for profit because Vermont utilities were not mandated to "retire" RECs. Utilities would then promote their energy as renewable, even though legally they had lost that right when they sold the RECs.

Ridgeline wind turbines are seen as an affront to regional identity and autonomy because they figure so prominently into the profits and electric needs of other regions. A recurring theme of opposition decries the injustice of siting turbines and solar panels in the poorest parts of Northern New England while exporting the RECs to affluent east coast cities. Ralph put it this way: "We're becoming the West Virginia of wind because we have this resource that gets developed here, and Massachusetts buys it . . . well . . . they use the energy credits. They buy the renewable attributes of it to satisfy the requirement in Massachusetts." Ralph's collaborator Jim added, "As far as producing clean energy, this project, it doesn't help our cleanliness at all because we sold that away."

Many early wind opponents in Vermont referenced the state's reliance on tourism and its mountains' appeal to visitors. As turbines have appeared atop the region's ridgelines, it seems doubtful that the appeal to visitors has waned, as many once feared, but as chapter 3 describes, these concerns tap into much deeper tensions between urban and rural equity and prosperity. Because wind turbines promise few permanent local jobs, opponents fail to see them as local economic development despite public officials' celebrating them as such.

In sum, wind opponents resist ridgeline turbines for multiple reasons. Despite agreeing that climate change is a pressing problem, the environmental community is divided over the issue. I will detail in later chapters how divisions spring from contrasting orientations to local versus global ideas about justice, to balancing urban energy needs versus rural autonomy, and to the belief that inadequate weight is given to greenhouse gas emissions.

Procedural Antiwind Arguments. My informant Christine is a pasta-throwing wizard. She holds boundless knowledge about electricity systems, renewable technologies, and energy policies in general. Surpassing four hours, our interview will probably be the longest I conduct in my life. Although Christine could cover a wall with reasons why wind is poorly suited for the region's ridgelines, she started out as a fan. She described the first public wind meeting she attended and how excited she was to see her state pursue renewable energy: "There was a wind proposal . . . and I went and spoke at the siting board meeting in favor of it . . . saying, you know, look at where you are. . . . Come on! Wouldn't you rather have this than natural gas? Only one hundred ninety-seven feet tall then, a lot of people had seen them. I thought they were beautiful, you know, we knew if it was an isolated area, there would be no neighbors who are complaining." That last part, about neighbors, is the key to Christine's about-face on industrial wind. As I will describe, she has a number of complaints, but at the core of her involvement with the issue is her belief that ordinary citizens are getting run over in energy planning processes. One does not have to look far to see how this happens and how it affects people.

Northern New England's political geography predisposes the region to wind conflict between neighbors. Land holdings are relatively small parcels, the mountainous terrain puts turbines on display for miles, and territory is divided into square-shaped towns. For those unfamiliar with New England, this means that very few residents live within "unincorporated" territory, with the exception of a few tiny parcels in Vermont and New Hampshire called "gores" and the vast but sparsely populated northern tier of Maine. The municipalities that host wind turbines receive property taxes from them and sometimes other direct payments. These infusions may reduce residents' tax bills regardless of whether one can see or hear the turbines from their property. This means that receiving financial benefits depends on the municipality in which one happens to live, not necessarily how close one might live to wind turbines. For example, towns surrounding one of Vermont's largest installations receive "good neighbor" payments that are smaller than the payments made to the town in which the turbines sit, even though homes in neighboring towns may be closer to the turbines. These disparities are easy to exploit.

Before I spoke to any informants, I attended a public wind hearing in 2013 before the Maine Department of Environmental Protection (DEP) in the town of Canton's small firehouse. It was easy to locate because the fire trucks were parked outside and the doorway was overflowing with community members. Canton is located in Maine's central-western region, about halfway between the Atlantic Ocean and the Canadian border, at a bend in the Androscoggin River. Census figures estimate that fewer than one thousand residents live there.

The DEP representatives visiting from Augusta looked out of place in the sweltering firehouse. They wore office clothes and sat rigidly at a table in front of several rows of folding chairs. Patriot Renewables, a Massachusetts company, received approval to install eight grid-scale turbines along a prominent Canton ridgeline. The residents who registered to speak included concerned citizens, businesspeople, local firefighters, outside experts, and wind industry professionals. Still new to the issue, I felt as though I were at a hearing on fracking or some other extractive activity with a questionable health and safety record. I watched and listened as members of the public drew links between wind turbines and every risk to humans and wildlife fathomable.

What was most surprising about the Canton meeting, was the level of community division it revealed around lines drawn by financial influence. We know from research across the globe that shared ownership or distributed profits compel support for renewable energy projects.[15] In Canton, it became clear that wind companies had divided the community by selectively distributing resources and eliciting support from specific community members. After the hearing, a bystander explained to me discreetly that members of the fire department who testified on behalf of the project were promised new equipment if the project was completed. A local preschool teacher who spoke on behalf of the project had been the recipient of a small business grant funded by the developers. All-terrain vehicle groups were promised access to the new roads built on the ridgeline. It was odd that this bystander told me all this, because I did not ask him, and rural New England is not a place where strangers talk much to each other. I think that watching his neighbors endorse the project without acknowledging their connection to it was so galling that he had to tell someone, and I just happened to be sitting next to him.

I have since heard and read similar stories across the region, and they resonate with stories from across the world. It has become common among wind resisters to refer to industrial wind companies as "Big Wind," and several local newspaper editorialists have described the coalition of wind developers as the "renewable industrial complex." This is for good reason. The industry is well funded, organized, and experienced. While some opponents contend that wind developers target the most in need with meager promises, my observations suggest instead that developers operate more strategically. They build community support for their proposals through a playbook used in the fossil fuel industry: they invest in local supporters to the extent that it sways public opinion in their favor, and they do so with a keen eye on electoral proportions and their bottom line.

Like payments for mineral rights, payments for hosting wind turbines on private property are negotiated privately and can thus vary between neighbors, fostering animosity. I have been told that gag orders are common, so that neighbors who have been paid are subsequently forbidden from complaining about a wind development. This variability carries over to negotiations at the municipal level as well. In the affluent community of Grafton, in southern Vermont, Spanish energy giant Iberdrola promised cash payments to every voter in exchange for approving the company's twenty-one-turbine proposal. The proposal was resoundingly defeated by community vote, and the developers have respected voters' wishes, though such votes are nonbinding. It seemed that with promises of cash, Iberdrola had hoped to divide less affluent year-round residents who vote in Grafton from seasonal homeowners who vote elsewhere.

My informants from Vermont recounted for me how developers hired members of two large local families to host dinner parties to build word-of-mouth support for one wind project. As Ralph described the parties, company reps showed up "in Gucci shoes, handing out twenty dollar bills and Vienna sausages." I could not tell how much of that statement was accurate, as Ralph did not actually attend a dinner party. But historians note that personal touches like this are tried-and-true practices for energy companies.[16] Such tactics helped brand Lowell Vermont's Kingdom Community Wind with an air of local ownership that the word "community" implies but does not actually denote. The project is officially called "Community" wind, but it is not owned by the community. It is a joint venture on privately owned land between a rural electricity cooperative and Vermont's largest monopoly electric utility, which is a partial subsidiary of a multinational fossil fuel conglomerate.

In addition to feeling shut out by developers, Northern New England's wind resisters feel constrained by the legal architecture that limits the communities' authority to shape local energy infrastructure. Five of the six New England states, including Maine, New Hampshire, and Vermont, have constitutions that limit the authority of town governments to capacities expressly granted to them by their state. This precedent, known as Dillon's Rule, is named after a nineteenth-century Iowa Supreme Court justice and governs municipal rights to some degree in approximately forty states. Although the three Northern New England states have delegated authority to their municipalities to enact land use and zoning regulations, authority to assess matters of energy remain under the umbrella of states' public utility laws, which, as I explain in chapter 5, have a decidedly undemocratic history.

In Northern New England, where participation in local governance is deeply revered tradition, adhering to Dillon's Rule on matters of energy is disheartening to many. The citizens empowered to fill the region's hallowed halls of government each March on Town Meeting Day are limited in their ability to engage with the era's most pressing land use issue. After several years of conflict, Vermont enacted Act 174, which attempted to solicit local input by promising "substantial deference" to towns that draft an energy plan, but the law requires towns to undertake an expensive expert-driven planning process. Act 174 has left many wondering what "substantial deference" means when Dillon's Rule affords municipalities no right to plan for energy to begin with. One online commenter observed: "substantial deference is a meaningless phrase to the Public Service Board. Public service is also a meaningless phrase to them." Towns that do not submit a plan receive only "due consideration," which seems equally ambiguous.

Video of one public wind siting meeting illustrates the source of frustration. When a citizen complained that several questions submitted to developers years before had gone unanswered, the developers' attorney, with apparent sincerity, recommended that citizens refer their questions to their town's attorney, who could compile the questions and then submit them formally to the developers. The town's attorney then stood to support that proposition but indicated that questions should first be vetted through the town's panel of elected leaders, called a selectboard. This protocol is probably a matter of coordination and billing for attorneys' time, but to the residents in attendance, it appeared an exhausting procedural obstacle course. Assembled at a question-and-answer session about the project, citizens were told that some questions should be passed through the town board to two sets of attorneys.

Feeling excluded from energy decision-making is now common, much to the detriment of the swift post-carbon transition we need. American energy politics have never been particularly democratic or inclusive, and before ridgeline wind projects ignited passions in Northern New England, most ratepayers were content to leave energy decisions to their legislators. But those days are over. While awakening public interest may seem like a good thing for more democratic energy decisions, in Northern New

England, opposition to wind has undermined what was once a robust trust in state government. My informant Ralph is a former state official with decades of experience in environmental regulation. He is from outside the region, so he recognized that the participatory governance possible in small rural communities is special. Ralph told me, "Everyone here celebrated the value of small towns and the contribution of the small towns to the state culture . . . and so forth. Now we enter a realm with regard to energy development, especially renewable energy development in which the small towns are . . . basically, elbowed out of the way and essentially . . . declared irrelevant. The feelings and claims of citizens of small towns and the impacts on their lives are shoved to the side."

My informant Jim is a lifelong Vermonter who runs a small business but has deep roots in agriculture. He described losing his faith in state officials' ability to deliberate transparently and resist corporate influence. "Well, honestly when we got into this [ridgeline wind proposals], I thought that if you made a good case, the government would hear your side, and do the right thing. I'd always had a lot of respect for the way I had thought that government worked. Since then, I think that the internal backroom politics in the state are probably the dirtiest, sneakiest, corrupt bunch of crap that I've ever seen, putting us in line with the reputations that you hear for New Orleans and Providence, Rhode Island and Boston. It's absolutely corrupt and I have no faith in it whatsoever anymore."

Both Ralph and Jim have been vocal opponents of a nearby wind project. While they did not fight the legislation that first made ridgeline wind viable, they were active in trying to halt construction. Since the project's completion, they have tried to hold developers accountable to various operating and environmental regulations. This has required raising hundreds of thousands of dollars in order to pay for legal representation to secure the opportunity to hold "intervener" status with the PUC. Without lawyers, ordinary citizens are ill-equipped to navigate energy bureaucracy. The Vermont PUC, which is officially known as "quasi-judicial" and uses court-like procedures to receive testimony, is a panel of three political appointees with backgrounds in government and the energy industry. Ralph and Jim recounted to me that none of their interventions over several years had ever resulted in a favorable ruling by the PUC, whereas the rate at which the project advanced meant that the PUC granted repeated and substantial leeway to the developer, which was allowed to commence blasting before a significant property dispute was adjudicated.

It suffices to say that exclusion from the procedures associated with energy governance have instilled feelings of disillusionment and alienation. I met with Ralph and Jim at an austere whitewashed town hall packed snugly in a village near a bustling general store, a church, a town common, and a post office. We spoke in the town hall's large open room with a stage on one end that served multiple functions, including hosting the annual town meeting and occasional ping-pong games. There were flyers and public service announcements stuck to the walls and clutter that felt comfortable and familiar. I had spent much of my early life in similar places because my dad worked in town government for decades. Through extracurricular activities in my high school years, I had seen state and local government functioning at the same human scale that Ralph and Jim were describing, and so I understood the loss and disillusionment in their voices when they described their foray into energy politics. In plenty of places, people do not expect much from their elected officials, but rural New England politics have a strong tradition of open dialogue and neighborly trust, which make one think that their voice matters. To think that those traditions are eroding could be a jarring existential turning point, and I could hear that in Ralph and Jim's voices.

No NIMBYs, Please: What Comes of Dissent

The vast majority of news coverage and virtually all of my informal interactions over the course of my fieldwork categorize wind opposition as NIMBYism, with "NIMBY" standing for "not in my back yard." The nuanced technical and procedural arguments that wind resisters make are undercut by the assumption that they are just bitter about developments near their home that they do not like. This is a problem both attributable to wind resisters' inability to message the public, and to the public's inclination to accept developers' narratives over those advanced by NIMBYs. The consequence is a very limited public grasp on renewable energy.

It is difficult to see Northern New England's fragmented wind opposition as a cohesive social movement, but it is useful to see opponents' lack of cohesion through the lens of social movements research. In our interviews, I asked explicitly how wind opponents orient against wind personally and how they think they should orient collectively to garner public support. Even among close collaborators, my questions elicited widely varying responses. One informant told me, "I don't feel like that's my strong point. I don't know if it's any of our strong points." Given how thinly stretched my informants' time was, and considering that all of their resources went to attorneys, my informants seemed to rely on trust in one another's competence to message the public as they saw fit. This decentralized strategy is partly a product of how differently the public relates to the wind issue. Property rights advocates, cost-focused ratepayers, climate-concerned voters, libertarian localists—all of them are motivated by different messages. Focusing too much on cohesive messaging may seem futile.

Wind opponents' passive approach to public relations stands in marked contrast to more organized social movements. Social movement research finds that groups negotiate a frame and then posture strategically so that the media accurately represents the groups' chosen frame.[17] In contrast, wind opponents with whom I spoke seemed to let the media cover the opposition as they wished, often resulting in media framing that treated wind opposition as individuals working in parallel rather than gathered collectively in a huddle. This framing was not necessarily a bad thing, because some locals saw any organized group as somehow being influenced by outside money, which is typical of rural New England's heightened mistrust of outsiders.

Unsurprisingly, several informants lamented the media's superficial reporting on matters of energy, and in particular, the media's weak coverage of wind opponents' structural energy critiques but strong coverage of opponents' critiques of turbines as structures. My informants see reporting in small and local papers as more respectful to nuance, but they believe the region's largest newspapers cater to the interests of wind developers, which portrays opposition as NIMBYism.

For the pro-wind public in the region, NIMBYs are a nuisance. Because NIMBYs are motivated by development in a specific place, society has been conditioned to ignore them. By invoking the word backyard, the NIMBY label implicates a place-bias clouding one's judgment, effectively categorizing development opponents foremost as selfish.[18] The label signals that one's position should be excluded to make room for unbiased, rational stakeholders, which affirms that developers, with their data, cost projections, and experience, are the trustworthy ones in the room. In the energy sector, this is a well-established tactic.

To illustrate, one study on resistance to "smart" electricity meters in Washington State found that energy company officials dismissed the technology's opponents as "black helicopter" conspiracy theorists.[19] Similarly, interviews conducted with wind developers in the United Kingdom found that "developers consider members of the public to be an 'ever-present danger' who could at any moment act to obstruct their proposals, a situation that required the provision of appropriate information."[20] Categorizing people as on the fringe or as the bearers of inaccurate information limits the likelihood that members of the public will parse out opposition arguments for themselves, to ascertain which threads are empirical and communal, as opposed to personal. This categorization of opponents works to the advantage of energy developers, and they know it.

I saw this approach at work in my field research. At public meetings, wind developers would make statements like, "I'm certainly aware of what you're seeing on the internet" as a way of discounting the public's knowledge as untrustworthy. At another meeting, a wind representative remarked, "If folks are coming to us with legitimate questions, we want to work with them," making it clear that certain "illegitimate" antiwind perspectives were unwelcome. For example, on the issue of declining property values, I watched a developer refuse to acknowledge a question that an audience member had asked: "Yes or no? Are you committed to helping us if our homes are rendered unlivable or our property values decline? Are you going to help us?" The developer said nothing, making no concession or admission, and thus leaving open for interpretation what the silence meant. While some observers might have interpreted it as evasion, those familiar with the industry's denial that turbines lower property values might interpret the developer's silence as a refusal to engage with irrational people asking ignorant questions. Public meetings about wind development seemed to be choreographed in this way, with tactics to separate the "rational" from the "hysterical."

"NIMBY" has become a powerful label in part because there is no legal precedent for defending one's right to unfettered natural space that is equivalent to one's right to profit. Put differently, our laws and legal system are better equipped to defend individuals' and businesses' right to make money than to enforce boundaries when profit seeking infringes upon others' rights, such as the rights to clean air and water, quiet nights, and dark skies. Thus, asking neighbors to sacrifice for profit is routinely regarded as a more legitimate request than asking businesses to sacrifice profits. Insofar as labeling or categorizing energy opponents invalidates their passionate place-based arguments, it is allowing energy developers to stifle important energy conversations. We need those conversations, and we need them to be inclusive of matters of equity and justice, because there is much that goes unheard.

Of utmost importance is the revelation that many, though not all, wind opponents are themselves motivated by global concerns.

In summary, on the surface of wind opposition are superficial and often aesthetic arguments about how wind turbines impact their local surroundings. Those arguments lend support to the developer's perspective that resisters are letting emotions stand in the way of progress. Focusing on surface arguments distracts attention from other arguments, which, in the case of North American wind resistance, comprise an array of critiques targeting entire energy systems, patterns of development, and modes of exclusion.

For decades now, social scientists have rebuked the oversimplified use of NIMBYism[21] and have produced volumes on the vital importance of participation and inclusion in natural resource decision-making.[22] We know that mounting a defense of one's backyard is not necessarily a selfish political stance. It is in fact possible, if not probable, that individuals acting in defense of a locality are motivated by their love of a community and that after educating themselves about development they become more active and engaged citizens. Rather than invalidating one's concerns, passionate defense of one's backyard can inspire deeper knowledge that benefits the public. For example, research conducted across Canada found that distrust in energy companies was often associated with trust in scientists and scholars as well with greater engagement in energy issues.[23]

Words like "ownership," "personal investment," and "stakeholder engagement" round out a vast evidence-based literature of best practices to ensure that environmental decisions, even when contentious, are at least made using processes that people trust. A literature on energy decision-making has also evolved, arriving at similar calls for making energy decisions as open and inclusive as possible.[24] For example, in an assessment of Germany's renewable energy successes, researchers found that opposition eventually led to more inclusive democratic decision-making processes, concluding, "the more that various local interest groups and individuals are involved in planning and implementation processes, the more open discussions are; the more responsibility local actors are given in the search for solutions and the balancing of alternatives, the more likely it is that they will perceive decision making processes as fair."[25] Despite all of this experience and analysis on open decision-making processes, in one of America's most politically progressive regions, people feel shut out from planning their communities' energy futures.

The impact of Northern New England's opaque regulatory environment is a public record of wind contention that would suggest to outsiders that aggrieved neighbors will not stop griping, even after turbines are built. Because development and regulatory processes prioritize surface-level complaints and offer no real venue for technical complaints, the NIMBY narrative spreads with every news story about actions to enforce sound standards and storm runoff, the only arguments for which opponents have grounds to petition. The nuanced and well-researched arguments that opponents have amassed in opposition to ridgeline wind are lost in the formalities and rubrics imposed by the regulatory process, and public knowledge of wind energy's regional drawbacks remains low.

Antiwind environmentalists appear to have changed few of their colleagues' minds on the issue. Some local antiwind groups have used their experience to promote thoughtful collaborations and alliances with groups in neighboring states. However, most just align with antiwind groups opposing projects around the world via the internet, which reinforces the NIMBY narrative by amplifying the pasta-throwing strategy. Meanwhile, wind advocates, including nationally renowned environmental luminaries, maintain steadfastly that wind-generated electricity constitutes an important first step toward reducing carbon emissions. These advocates believe that local ecological concerns should not trump the global imperative to mitigate climate change, which is a common narrative propagated by social research throughout the world.[26]

Nonetheless, strong support from most of the environmental community has not been enough to maintain the same momentum in the wind industry. The pressures created by antiwind groups through negative publicity, legal challenges, and election outcomes have made turbine development more difficult globally. Since I began my research in 2013, Vermont overhauled its energy law and adopted stronger sound standards for new turbines that developers claimed would stifle new wind proposals. In 2021, Vermont and Hampshire have no new wind projects under review. Several projects continue to advance in Maine. Both Maine and New Hampshire have denied projects in scenic areas on aesthetic grounds.

Conclusion

Environmental sociologists Freudenberg and Pastor called for research that moves beyond blaming or understanding development's victims, to understanding "the broader systems that creates victims (and victimization) in the first place."[27] This is the approach I follow through this book. I do not frame wind opponents as victims; rather I understand them as part of a society that experiences collective victimization by an energy system following an antiquated mandate for cheap, abundant, and profitable electrons. If wind turbines are important parts of a renewable energy future, then I agree with many social scientists before me that we must seek to better understand the reasons behind antiwind sentiment. With that said, if we demand a renewable energy future that prioritizes climate change mitigation, we must open ourselves to the possibility that antiwind sentiment holds useful insights to overhaul antiquated and ecologically damaging energy systems.

Wind opposition is more layered than energy narratives presented in a news media that broadcasts sound bites about one particular complaint in one particular place or event. Opponents use an arsenal of arguments that work differently. Technical arguments approach wind turbines as an imperfect technology, contending that putting them on remote ridgelines does little to mitigate climate change. Opponents are also concerned about procedural transparency in determinations of where projects are sited, distribution of particular projects' costs and benefits, and developers' and financiers' trustworthiness.[28] After going unheard, opponents critique the policy processes that exclude them. Looking more deeply into Northern New England's energy discourse, past the sound bites and pasta throwing, reveals that ordinary people are already thinking about energy more broadly. They just have few venues in which to share their thoughts.

In digging through the archives of regional newspapers' reporting on wind, I came across a letter to the editor that encapsulates the more complex sentiments that are often missing from the public conversation. The full-page letter was written in 2006, when ridgeline wind was first presented to Vermonters as a possibility. The letter outlines a regional environmental ethos that is worth probing.

The author, from southern Vermont, begins by disclosing that despite being a committed environmentalist, they could not stand behind wind energy as it was being proposed. They thought that the scale was all wrong, writing, "I suppose they have to be large in order for it to be worth it to the companies to build them. . . . But that's not the Vermont way. We've never been gung-ho for huge projects—whether they're gigantic Wal-Marts or large cities. . . . I would love to see every home with its own small wind turbines in the backyard, just as we now have many individual homes with solar collectors as part of their architecture."[29]

This writer represents an important environmental perspective not often featured in media accounts of antiwind sentiment. Most people in Northern New England have believed for decades that energy should be cleaner and something must be done to address climate change. But turbine technology is far more complex than most people realize, and the number of turbines required to make a meaningful impact on the region's relatively small carbon footprint is impractical, so why has the region committed so hardily and so rapidly to the technology, and at a such a scale? Turbines' size and the timing of their arrival suggest to this person that profits seem to be driving the phenomenon instead of the environment. Moreover, big turbines mean big resistance from neighbors. These are astute observations that I will expand on later.

Embedded in this perspective are assumptions and beliefs about place that I will also try to unpack throughout this book. The letter writer talks about the "Vermont way," which speaks to an exceptionalism wrapped up with our state pride. Northern New Englanders nurture a potent strain of the American ethos valorizing stoicism and self-reliance, while glossing over obvious contradictions. Hosting a nuclear facility was not on-brand for small, liberal Vermont, but we did it, and we still buy nuclear power from elsewhere. We have no shortage of suburban sprawl or big-box stores, and anyone who has visited a New England ski resort has seen that New Englanders have no problem deforesting mountains for ski trails and condominiums. The "Vermont way" is a myth, or at the very least, it is not what we think. In a region with so many ecological contradictions, it is easy to selectively remember our green achievements instead of our embarrassments and to prioritize our eco-branding at the expense of the actual environment.

Northern New England's mountains and ridgelines are important agents in this green branding. The mountains are totemic symbols of purity that offer both an iconic backdrop and a visual barrier to unsightly development going unchecked in the valleys. But the "Vermont way" is precisely why Northern New England embraced ridgeline wind so fully. As symbols of environmentalism, wind turbines are part of the same green branding that many people associate with the mountains themselves. Wind turbines were a seductively conspicuous energy option for a region proud of its environmental legacy. Note how the letter writer above would like to see wind turbines in people's yards—just not big ones on the mountains.

Perhaps the problem is that we think we need to see energy solutions visually? It seems plausible that the best solutions to our climate and energy problems cannot be seen at all, as is certainly true of efficiency. Perhaps we are fixating too much on symbolic visages of energy transition to stroke our do-gooder, individualist egos? Maybe our affinity for wind turbines as environmentalists is as superficial as the aesthetic arguments we so often criticize as NIMBYism?

There is also good reason to believe that we are being coached to accept a transition road map charted by the same industries that have set us on course with climate disaster. In addition to understanding why wind turbines are unpopular, an assessment of energy transition merits a critical assessment of what modern turbines are, how they are situated in social relationships, and how we experience them symbolically. In simpler terms, we need to better understand the ways that ridgeline wind has been packaged as a transition solution, which will require accounting for both who does the packaging and who designs the transition. Ultimately, this effort requires us to become better informed consumers of electrons and more responsible participants in energy planning.

Though wind turbines have become resonant symbols of energy transition, the climate crisis demands energy transformation. Each of the following chapters uses lenses from environmental sociology to unpack, problematize, and contextualize how ridgeline wind fits within an energy transition prioritizing the status quo instead of the transformation that climate research demands. While wind opposition in Northern New England may resemble entitlement, it can also be seen as a deep-seated resistance to status quo nodes of political power. I describe in the following chapter how the seeds of wind resistance take root in an enduring tension over how cities dictate rural landscapes. I show how this urban-rural tension impacts not just how people feel about ridgeline wind development but also how they make sense of larger trends in renewable electricity.

3

For the Love of Mountains
••••••••••••••••••••

The Green Politics of Place

> I have, growing up a real flatlander,
> become emotionally and spiritually
> attached to these landforms called
> mountains.
> —Ralph

As fourth graders at Weathersfield Middle School, we learned in local history class about a farmer in our hometown who torched his house and killed himself rather than surrender his land for the construction of Interstate 91. His name was Romaine Tenney, and the story of his suicide would later inspire me to write a college paper exploring social change brought by Vermont's interstate highways. Writing that paper, I learned that decades before the interstate arrived, lawmakers debated paving a parkway along Vermont's mountainous spine, similar to the one in the Blue Ridge Mountains of Virginia and North Carolina.[1] Because the Green Mountain Parkway was never built, when the interstates came to Vermont in the 1960s, they showcased stunning panoramas of pristinely forested mountains. Thanks to those panoramic views, the region's mountains became more highly prized amenities. Our mountains are visual anchors to a tourism economy relying on visitors from the rest of the Northeast, locally known as "flatlanders," and the mountains are spiritual anchors for residents.

I can remember when New England rallied to protect its mountains from acid rain in the 1980s. The slow creep of defoliation advanced across the ridgelines year after year, leaving brown dead patches. I remember that environmental groups publicized before-and-after photo arrays documenting the creeping demise of our trees, and I remember learning that effluent from Midwest smokestacks was responsible. I was young, but I was horrified by the prospect of our forests dying off. Organizing and education campaigns encouraged the public to lobby for amendments to the Clean Air Act. The amendments passed, our trees were saved, and our shared investment in the region's iconic landscape was further cemented.

Reflecting on threats like paved ridgeline roads and slow deforestation, the uproar over wind turbines seems frivolous. Displacing coal from the electricity grid can only help maintain forest health. The infrastructure needed to install turbines on mountaintops seems no more harmful than that required to carve ski resorts out of wilderness. It is tempting to ponder whether wind opponents maintain an illogical hostility for wind turbines or an illogical affinity for mountains. But this line of thinking oversimplifies the situation.

All of the anger and disillusionment that I have described thus far is not happening in a rural region that feels suddenly invaded by industrial machines. Not only do wind opponents follow in a long line of impassioned mountain protectors, but wind development also follows in a long line of outside interests upending local autonomy and reshaping local ecology. The mountains are symbols of regional identity and autonomy. Go after the mountains, and you stir a deep pot of resentment for an economic logic from "away."

Whether lumber, minerals, or electricity, extracting a resource from a rural community and shipping it to a city where shareholders count the profits is a familiar economic arrangement. Energy in New England does not perfectly fit this description, but wind development does follow a similar pattern. Considering Northern New England's history as a rural periphery, and considering where wind energy will be sold and whom it will benefit, it becomes easy to see where the "us versus them" energy rhetoric originates and how belief in the mythical "Vermont way" proliferates. This chapter describes how ridgeline wind agitates Northern New England's deeply ingrained sense that it is unfair to be beholden to outsiders' vision for the landscape, encouraging deeper thinking about what is fair in energy.

Rural by Design

Understanding the backlash against wind requires first understanding Northern New England as a distinct place shaped by an iconic landscape, by the people who have flocked there, and by the political culture that evolved as a result.

Maine, New Hampshire, and Vermont have always been more sparsely settled than other states in the Northeast. The region is notable for its lack of major cities, the modest size of its largest towns, and its status as a vacation destination. When I taught at Bowdoin College in Maine, even students who had grown up in the region struggled to identify rural communities on a blank map that I printed for them, an exercise that I would repeat every semester. As I recall, most students failed to identify by name more than a dozen of the region's cities and towns, even when working in groups.

Northern New England's sparse population can be traced back to deliberate policy actions in service to the region's cities. Before the twentieth century, only a few slivers of the three states were protected natural lands. These were resorts for East Coast elites, on lakeshores, in mountains, and at mineral springs.[2] Much of the mountainous region had experienced clear-cut logging followed by grazing, which erased brooks and streams from the land and altered the drainage systems feeding the Northeast's waterways.[3] Southern New England experienced flooding as a result of the denuded hills and mountains, prompting federal legislation to protect large swaths of the Northeast from future development. With the Weeks Act of 1911, the U.S. Congress laid the ground for the federal purchase of vital stream habitat in the forests of Vermont and New Hampshire. The White Mountain National Forest was acquired in 1913 and the Green Mountain National Forest in 1928.[4] Today these forests account for major portions of New Hampshire and Vermont, respectively.

Although acquired to protect cities from flooding, national forests became central to regional identity and, eventually, to regional environmental politics. The Weeks Act was championed by the leaders of newly formed environmental groups, such as the Appalachian Mountain Club and the Society for the Protection of New Hampshire Forests. Its implementation reinforced Vermont and New Hampshire's formal efforts to craft identities rooted in forests, instead of hilly farmland. For example, the 1931 report of the Vermont Commission on Country Life advised that high-quality roads were needed, not for commerce, but to funnel to the state "seekers of health and beauty, drinking in from many landscapes the enchanted draught that leaves the gazer restless and unsatisfied until he can return." The Commission's report continues: "And so we mean to guard with jealous care the nobility and freshness of our scenery. Here is that wealth that never can be exhausted but by our own stupidity."[5] State leaders had commissioned this report to address sustained declines in agriculture and population that left the state economically stunted. The report's recommendations charted a course for tourism and recreation development that would further strengthen the emphasis placed on appealing to visitors.

Maine's status as a self-described destination dates to the same era. In fact, Maine boasts the nation's longest-lasting license plate slogan, which advertises the state's destination status with the word "Vacationland."[6] Though Maine's coast is the state's most heavily visited region, Maine's inland mountains and lakes are as glorious and popular as New Hampshire's and Vermont's, albeit harder to access.

Regional tourism benefited next from federal policies of the Depression era. As part of President Franklin Roosevelt's New Deal, the Civilian Conservation Corps (CCC) employed tens of thousands of laborers to clear mountain roads, cut hiking and ski trails, create reservoirs, and erect park buildings.[7] This infrastructure formed the backbone of the region's nature-based tourism in more than one hundred state parks, ski hills, national forests, and federal recreation areas, including the final stretches of the Appalachian Trail. From the 1960s to the 1980s, ski resort growth appealed to winter tourists, spawning massive real estate booms across the mountains of all three states.[8] Skiing expanded the tourism industry at a time when farms continued to close or consolidate and manufacturing slowly succumbed to globalization.

When the Tourists Stayed: Green Voters, Policy, Energy

Northern New England's appeal to guests has long inspired many to move to the region permanently. Though much of the United States experienced its first "rural renaissance" in the 1970s, Northern New England had a long and steady history of urban in-migration.[9] After European settlers violently displaced the region's indigenous peoples, Northern New England became one of the United States' original rural destinations, as the region's cooler climate, cleaner waters, and thicker forests look and feel very different from the populous areas to the south.[10] Thus, the region's social fabric has been continuously reshaped by outsiders attracted to its wild terrain and bucolic small towns, and while homogenous by race and ethnicity, Northern New England weaves together successive waves of newcomers.

The first outsiders were affluent urban residents seeking summer homes, inspiration for their art, or a permanent escape from urban ills. Northern New England is, thus, a region that harbors stark social inequalities, where class divisions are deeply ingrained and where sentiments about rural and urban cultures and lifestyles have hardened for centuries. The countryside is home to prestigious private colleges and boarding schools, opulent resorts, artist's colonies, and summer "cottage" communities. The region's unique juxtaposition of local aristocracy and abject poverty has inspired novelists, filmmakers, and television producers. It is through this regionally nuanced experience of in-migration and class inequality that the region's energy politics must be understood.

Three waves of in-migration roughly corresponding to decades are relevant to the region's reception of wind energy projects. In the 1970s, young baby boomers were enticed by the region's cheap farmland and mountain vistas at a time when youth were growing disenchanted with mainstream society and when finding work in American cities was a challenge. These young people took up skiing, lived off the land, raised families in drafty houses heated with wood, and started small farms and businesses that helped redefine the region's culture.[11] In the 1980s, as eastern cities faced high crime rates and declining financial markets, Northern New England enticed newcomers to the region with new work opportunities. It was one of the few rural regions in America to experience growth from cities in the 1980s.[12] Always a staple of the region, the tourism industry expanded and professionalized rapidly. Ski resorts grew their footprints and corporate operations, expanding into hotel and real estate development and summer attractions.[13] Spilling over from southern New England, the region's small cities developed technology-driven economies, bringing skilled managers and engineers to the region. Meanwhile, traditional rural economic sectors continued their declines, and the rural brain drain accelerated. Starting in the 1990s the region began attracting lifestyle migrants of all types. Retirees seeking idyllic settings for their golden years, aspiring farmers, and urban outdoor enthusiasts were untethered thanks to the ease of telecommuting, and they found many affordable communities from which to choose.

Generally, "newcomers" are more likely to be at or near retirement age, more educated, and more affluent than the average resident born in the region. In relative terms, their wealth stands in stark contrast to that of a small working class and scattered pockets of rural poverty. Newcomers help make the region the oldest and most racially homogenous corner of America, as well as one of the country's most politically liberal regions.

What old-timers and newcomers share is a near universal appreciation for the rural landscape that anchors them to the region. Over time, identities in the three states have internalized the unique and economically important qualities of the region's natural landscapes, and it has become second nature for policy makers to closely guard those qualities. One study from New Hampshire estimates that 76 percent of the state's residents participate in some form of outdoor recreation.[14] Compared with other parts of the United States, a relatively large proportion of the region's residents have moved there from elsewhere, exercising a choice to live in a region far richer in environmental splendor than in economic opportunity. Newcomers grew to celebrate and emulate the stoic, independent, and thrifty archetype of New England farm families; thus, Northern New England's politicians have long favored economic development that works in synergy with environmentalism and preservation. The region was primed for growth in the renewables sector by a long-standing interest in environmentally conscious development and an even longer-standing interest in maintaining natural landscapes—even as Maine, New Hampshire, and Vermont have each elected both liberal and conservative governors over the past decades.[15]

To summarize, ridgeline wind development is only the most recent example of green policy in the region. A century of landscape preservation has ingrained a reverence for mountains and promoted an economy and regional identity predicated upon their natural appeal.[16] The landscape has shaped environmental politics through its influence on voters and lawmakers and through the region's reliance on tourism. Wind developers draw from this legacy. Wind opponents do too, as well as from a legacy of activism resisting alterations to the landscape.

Northern New England Green Population, Green Policy Time Line	
1876	- Appalachian Mountain Club founded in Boston
1897	- First AMC hut built in New Hampshire
1901	- Society for the Protection of New Hampshire Forests founded
1910	- Green Mountain Club founded in Vermont
1911	- Weeks Act allocates federal funds to preserve NE headwaters
1918	- White Mountain National Forest acquired under Weeks Act
1919	- Acadia National Park opens
1930	- The Long Trail, spanning length of Vermont, completed
1931	- Vermont Commission on Country Life publishes report
1931	- Maine's Baxter State Park founded
1931	- First winter tourist train arrives in New Hampshire
1932	- Green Mountain National Forest acquired under Weeks Act
1934	- World's first ski-tow installed, Woodstock, Vermont
1935	- Green Mountain Parkway defeated
1936	- Maine adopts "Vacationland" license plate
1950s	- Vermont builds mountain roads, leases 8,500 acres to ski resorts
1968	- Vermont enacts billboard ban
1970	- 24% in region born out of state
1970	- Vermont enacts ACT 250 (environmental review)
1977	- Maine enacts billboard ban
1980	- 33% in region born out of state
1988	- Maine enacts Natural Resources Protection Act
1988	- New Hampshire enacts Rivers Management & Protection Program
1990	- 39% in region born out of state
1995	- Vermont Sustainable Jobs Coalition forms
1997	- Maine establishes Small Enterprise Growth Fund
2003	- Maine Lead by Example Act to track greenhouse gas emissions
2007	- New Hampshire enacts Climate Change Action Plan
2008	- New Hampshire Renewable Energy Generation Incentive Program
2012	- Vermont Passes Universal Recycling Law
2014	- Maine, Vermont pass labeling law for genetically modified foods
2020	- Vermont enacts Global Warming Solutions Act

FIG 3.1 Time line of regional environmental policies

Mapping the New Electricity Landscape

Detailed studies now estimate the enormous scale at which renewable electricity will reshape rural areas. Rural space will not only host solar arrays and wind farms on behalf of the world's cities but also provide the various mineral components of which solar panels and wind turbines are built. Wind opposition in Northern New England nests within this global realization that the green ambition of urban electricity customers will create a large footprint in the countryside, beyond rings of suburbs where people live in expensive homes and enjoy high-paying jobs. Many of the region's rural communities are losing residents and gaining open space, making them even more attractive locations for renewable energy. Meanwhile, many wind opponents' experience with the wind industry tells them that they can no longer participate in land use decisions, once the domain of deliberative community dialogue.

An electricity grid, with immense power plants and millions of miles of cables, poles, and substations, offers an effective conceptual parallel to the networks of political and economic power that govern the grid's design and regulation. The actors who design, build, manage, and regulate the grid rely on relationships in the hinterlands to maintain control. The way the grid generates and allots electrons to various end users mirrors the way wealth and political clout shift across space. This analogy is useful for understanding wind opposition because contemporary renewables policies rework the geography of both the "power" that refers to electricity and the "power" that refers to politics and economics.

Just as political and economic power clusters in cities, electricity is normally generated in central locations, making those places essential to the functioning of the grid. Because electricity dissipates over space, generating facilities have historically located near customers, meaning in or near cities. In rural locations, infrastructure is sparser, older, and often suited for primarily residential demand. Rural locales were never as profitable for electric companies because of the costs associated with transmission, which is why so many parts of rural America receive electricity from cooperatives. (Things are different where hydroelectric dams and electricity-intensive industries such as mines locate. In those rural places, opportunities for profit exist, and private electric companies have thrived.) Renewable energy is redrawing this map, making places long forgotten about by the electricity sector suddenly appealing.

While social scientists have penned volumes explaining how one's proximity to wind turbines informs the likelihood that one will oppose them, relatively little has been written about the distance electrons travel from turbine to light switch. Of course, this distance is conceptual, because individual electrons cannot be tracked; but what can be tracked is the exchange of funds between those producing electrons and those paying for them. In the age of renewable energy trading, that distance is large and growing. In modern electricity markets, one must pay attention not only to where the region's wind energy is being sold but also to the location of those who are brokering and profiting from the exchange.

The players in in Northern New England's electricity markets vary in a few important ways. The area is served by a number of rural electric cooperatives, which are owned by ratepayers. Cooperatives were formed in the 1930s, when electricity businesses focused their attention on cities, where population density promised large profits. Additionally, the region is home to a handful of municipally owned utilities, which tend to serve more densely settled places. More than 80 percent of customers in the region are served by investor-owned utilities (IOUs), which are owned by shareholders who receive regular dividends from the IOUs' profits.

All four of the region's IOUs have been acquired by out-of-state parent companies. Maine's largest utility, along with the bulk of its transmission infrastructure, is a subsidiary of Spanish conglomerate Iberdrola. The state's second largest IOU and many large hydroelectric projects were a subsidiary of the maritime Canadian company Emera, until it sold its Maine assets in 2020 to ENMAX of Calgary, Alberta. New Hampshire's largest IOU was purchased by New England's largest electric utility, based in Connecticut, now operating under the name Eversource. Vermont saw its two largest utilities merge, and they are now subsumed under the Quebec company Énergir, which is owned in part by the Canadian energy conglomerate Enbridge, known for its investments in gas pipelines. (Some Vermonters fear that Enbridge is posturing to eventually increase the volume of fossil fuels passing through its pipeline in Vermont.)[17] In other words, within in a decade, two of the region's largest energy companies have become wholly owned subsidiaries of Canadian energy conglomerates, and a third is owned by a European company. This web of external ownership combined with the wind industry's new network of players maps a new hierarchy of place, which is essential for understanding why wind development across the region became so contentious so quickly.

Electricity Policy History

1932 - Eight electricity holding companies serve 73% of U.S. customers

1935 - Public Utilities Holding Company Act - Established monopolies to solve corruption

1978 - Public Utilities Regulatory Policies Act - Requires purchases from small generators

1992 - National Energy Policy Act - Makes it easier for generators to enter the market

1996 - Federal Order 888 - Opens access to nondiscriminatory transmission services

1997 - ISO-NE assumes New England grid management role from NEPOOL

1998 - New Hampshire deregulates electricity market

2000 - Maine deregulates electricity market

Maine Investor-Owned Utility Consolidation (90% of customers)

2000 - CMP sold to Energy East (parent of New York State Electric & Gas)

2001 - Emera (Nova Scotia, Canada) acquires Bangor Hydro as Emera-Maine

2008 - Energy East sold to Iberdrola (Spain)

2010 - Emera-Maine acquires Maine Public Service

2015 - Iberdrola USA consolidates new acquisitions under Avangrid

2020 - Emera-Maine sells to ENMAX (Alberta, Canada)

New Hampshire Investor-Owned Utility Consolidation (70% of customers)

1992 - Public Service of New Hampshire joins Northeast Utilities (Hartford, CT)

2012 - Northeast Utilities acquires NSTAR (Boston, MA)

2016 - Northeast Utilities becomes Eversource

Vermont Investor-Owned Utility Consolidation (80% of customers)

1986 - Gaz Métro (Quebec, Canada) acquires Vermont Gas Systems

1997 - Hydro-Québec acquires 42% stake in Noverco, parent of Gaz Métro

2010 - Gaz Métro becomes 71% holding of Noverco (a subsidiary of Trencap [61.11%] and Enbridge [38.9%]), and 29% publicly traded Valener

2007 - Gaz Métro acquires Green Mountain Power (GMP)

2011 - Gaz Métro acquires Central Vermont Public Service (CVPS)

2012 - CVPS merges with GMP

2014 - GMP becomes B Corp, "using the power of business to alleviate poverty, address climate changes, and build strong local communities and great places to work."

2017 - Gaz Métro becomes Énergir

2019 - Vermont regulators approve privatization of Énergir parent Valener

FIG 3.2 Time line of electricity policies and regional consolidations

Wind turbines account for only one part of Northern New England's new energy politics. The second heated issue playing out in hearing rooms, newspapers, yard signs, and bumper stickers concerns electricity transmission projects transecting the region to connect Quebec's enormous hydroelectric dams with customers in Massachusetts. With the phaseout of nuclear, coal, and oil power plants, East Coast cities will need power from Canada's enormous reserves of hydroelectricity. Capacity from dams on just one Quebec river coming online in 2021 will generate enough power for 1.5 million homes.[18] Contracts with utilities in Boston and New York necessitate major transmission projects to deliver electrons across the U.S.-Canada border and through Northern New England and Upstate New York. The first of these projects, the Champlain Hudson Power Express, will see construction of underground cables following a route beneath Lake Champlain and the Hudson River.

To deliver hydropower from Quebec to Massachusetts, utility companies have proposed several routes through Maine, New Hampshire and Vermont. The most contentious route was the "Northern Pass," charted through New Hampshire's storied White Mountains, which failed to gain approval in 2018. The route stirred vehement backlash across the state, resonating with year-round residents and seasonal homeowners alike. Two routes have been proposed through Vermont with far less contention. They were abandoned around the time that Massachusetts utilities chose to explore a proposed route through Maine instead of New Hampshire. Immediate and vocal resistance to the "Maine Clean Power Connection" emerged-just as it did with the Northern Pass, though in early 2021, the project's approval seems imminent. Nowhere has regional solidarity proved stronger than in resistance to these corridor projects.

A 2011 federal rule makes corridor projects transecting but effectively bypassing rural communities even more contentious by forcing those communities to pay a bigger share of them. The Federal Energy Regulatory Commission's (FERC) Order 1,000 requires states that once paid for transmission only in proportion to their electricity use to subsidize neighboring states' new power lines. For example, Vermont, accounting for 4 percent of New England's electricity use, must split 70 percent of the costs of Massachusetts' new transmission needs with the other four New England states. This means corridor projects entail more than aesthetic costs or a symbolic slight. Under Order 1,000, rural ratepayers are expected to

FIG 3.3 Sign opposing the Northern Pass in Coos County, New Hampshire (Photo courtesy of Shaun A. Golding)

foot the bill for things for which they fought against and may receive no immediate benefits.

The changing web of ownership and management in the energy industry has only strengthened skepticism among Northern New England's wind opponents. New ownership structures undermine the credibility of local wind developers, and proposed corridor projects compound northern New Englanders' heightened sensitivity to manipulation. Residents have learned that electric companies are bigger and more removed than the days when communities paid their bills locally. Chapter 5 explores these dynamics in greater depth.

Ruralism, Tourism, and Wind

Several social scientists have examined the ways that communities invoke place attachment to welcome or resist renewable energy.[19] We know that local groups frequently disagree about whether a renewable technology is a match for their landscape and that in making that assessment people can

FIG 3.4 Sign opposing Central Maine Power's corridor project, Cumberland County, Maine (Photo courtesy of Shaun A. Golding)

rely on different orientations to global problems. When it comes to energy, social scientists have highlighted the irrational underpinnings of rural exceptionalism—the premise that conventional ideas about what "rural" is make it difficult to justify any rural development that does not fit our ideas. Environmental scholar Roopali Phadke's description of resistance to wind turbines in Nevada suggests that industrial technologies installed in rural landscapes become jarring distractions from expectations of what people think rural communities *should* look like.[20] Phadke draws on the work of geographer Michael Woods, who describes the tendency to defend rural landscapes from threats that seem out of character as originating from a perspective termed "aspirational ruralism."[21] But this would be an oversimplification of the dynamics in Northern New England, where aspirational ruralism better explains those who *want* to see wind turbines on their mountains than those who do not.

Perhaps for some recent newcomers, wind turbines seem like jarring additions to the mountains. But the region has appealed to urbanites for so long and so consistently that many so called "newcomers" have lived there for decades. Many who relocated were drawn to the region not by a postcard backdrop, but by a productive landscape allowing them to carve out a sustainable lifestyle. I was struck by this distinction when a few of my informants brought me to their houses. Christine, for example, lives nestled in a valley down a steep driveway off a winding dirt road. She moved to the region in the 1990s after a childhood spent bouncing around the East Coast. Her home sits along a stream, in a patch of bright green grass fertilized by roving chickens. Her back window frames a view of a large solar panel that provides all of her electricity. She served me what I think was my first solar-powered cappuccino. Christine is a well-rooted "newcomer" who represents countless educated professionals and stalwart contributors to the region's environmental community. Many are utterly thrilled to see wind turbines on their mountaintops, like Christine was at first. Of course, some "newcomers" want trophy homes with panoramic views and no trace of civilization— I have read about them in newspapers. But many residents are overjoyed to see icons of renewable energy through their windows. That is progressive New England's particular aspirational ruralism.

In light of this diversity of perspectives, we must acknowledge that wind opposition takes root not in an idyllic image of rurality but in an ingrained, power-laden process that created a particular rurality in Northern New England. The ways in which the region's residents assess wind projects' transparency, equity, and trustworthiness relates to residents' connections to landscape and how they think it fits within society's broader cultural and economic fabric. This assessment of "fit" has been influenced over time by environmental policies. While green energy dovetails with green politics for some, for others ridgeline wind amounts to a sudden script change for a region that needs to keep its mountains intact. Wind energy's proliferation marks the first time many residents have had to accept policies permitting alteration to the landscape, because for so long that landscape has been protected for tourists.

But I see tourism as a dog whistle for talking about inequality. Wind turbines have arrived at a time when tourism seems more important than ever for rural livelihoods. Scholars have labeled rural economies "post-productivist," when harvesting raw materials gives way to selling artisan crafts and culinary goods to visitors.[22] While the region is fortunate to welcome a captive tourist audience to sustain its array of artisanal products, jobs in the tourism and service sectors are notoriously low-paying and offer little upward mobility.[23] But mills and factories in the region continue to come and go, further strengthening dependence on tourism and service jobs.

The push for ridgeline wind turbines presents communities with conflicting messages about these communities' place in the region. At the same time that governors are fast-tracking renewable energy development, economic development experts are teaching rural communities to tailor amenities to urban expectations. This means that the infrastructure generating electricity for large cities out of state must fit in to landscapes that are more dependent than ever on aesthetics. A 2005 op-ed titled "Big Wind, Small State" appeared in Vermont's largest newspaper outlining what was at stake on the eve of major turbine developments. "This is the state that rejected billboards on roadsides, defeated the Green Mountain Parkway and put tough environmental regulations in place with Act 250. We cannot lose sight of Vermont's distinctive place in the world with its open spaces and gorgeous vistas. It is up to us to continue the legacy. Real jobs, real lives depend on it."[24] This statement should remind us that that even concerns about property values and loss of tourism revenue are legitimate economic concerns that society should confront transparently if it wants to see renewable energy proliferate.

The region's dependence on tourists, who sometimes seem to flaunt their wealth and status, makes residents particularly sensitive to demands placed upon rural areas by urban interests. For many wind opponents, the discomfort with this urban-rural relationship is a part of their identity. They have been conditioned through marketing, or perhaps even formal training in the tourism industry, to know that the landscape sustains the region through recreation and not extraction. Now, the landscape is an energy factory and an energy throughway. While no evidence exists that wind turbines have ever hurt tourism revenues anywhere, the existential conflict between mountains and tourism has been a prominent strand in Northern New England's energy debate. In fact, the prospect of high-tension power lines in New Hampshire's White Mountains was ultimately enough to stop the Northern Pass, showing that with sufficient support, the region's tourism-focused identity can indeed be mobilized to halt energy development.

The finding from Phadke's study of Nevada that is most relevant in Northern New England is that rural communities' relationship to cities is seismically shifting under the region's modern renewables regime. Electricity is now dominated by powerful interests from increasingly farther away and the grid is managed by a regional authority in which more populous states exert greater authority. Electrons, or more precisely the Renewable Energy Certificates (RECs) representing electrons, are traded over great distances, meaning that the communities hosting renewable energy infrastructure may actually see their carbon footprints increase. All of these uncomfortable realities about the region's wind development have made wind power a questionable policy for many Northern New Englanders, who have a strong connection to the landscape. Newcomers and lifelong residents alike make sacrifices in order to remain in a landscape they love. They eschew the enticement of cities' higher wages and commercial landscapes and many have perfected frugal lifestyles and financial habits to compensate. So of course they respond with righteous indignation when cities expect them to accommodate the urban thirst for electricity and the eco-guilt that accompanies it.

Online comments sections overflow with observations about how cities waste energy in ways that stagger rural residents. Posters recount trips to cities where they encountered air-conditioned stores propping their doors open in the middle of summer and a skyline of empty office buildings illuminated at night. One poster laments, "Must we destroy Vermont so NYC can be lit up like a million Christmas trees 24/7?" In a region that prides itself on frugality, the geographic realities of local renewable energy markets are infuriating.

FIG 3.5 Sign protesting ridgeline wind turbines in Island Pond, Essex County, Vermont (Photo courtesy of Shaun A. Golding)

Mountains Divide the Environmental Community

Defending mountains has become a political liability for wind opponents because it conflicts with people trying to defend the earth from climate change. Shane, who when we spoke was a journalist focusing on the environment, sees the wind debate as a battle between varying strands of environmentalism: "So far I think the two sides of the debate are: 'Alright . . . we need to generate more renewable resources,' but the other side of the debate is saying 'hold on, let's do it in a more methodic way that accounts for the environment, our health, and the general beauty of the state." It's saying, 'let's step back a little bit and talk about the best way to do this.' But I think how we get there is a work in progress." Shane marvels that the most vocal parties in the wind controversy agree along axes that in other places would be deeply contentious; They all acknowledge human-caused global climate change and seek aggressive steps to reduce carbon emissions.

Christine shared with me her perspective on the division in the environmental community: "I never in a million years could imagine that I would see my former colleagues on ground water protection be the ones issuing the permits to allow the filling of class A1 headwaters above 2,100 feet. But they were." Like others I interviewed, Christine referenced Bill McKibben and Bernie Sanders, progressive luminaries whom interviewees once considered friends and allies. Sanders and McKibben are outspoken supporters for wind energy, and despite a history of healthy professional relationships, have become unsympathetic to Vermont's wind opponents. When reflecting on this fracture in the state's environmental community, several of my informants conjured the solemn tones of regret we associate with divorce. Karen told me, "The progressive environmentalists . . . the Bill McKibben followers, I feel like that's the hardest group to address. The minute you say you're opposed to wind, they think you're a redneck climate change denier."

Jacqui, whom I met through Ralph and Jim, says that the wind issue is "tearing the environmental community apart." While pro-wind environmentalists award less credibility to antiwind environmentalists, for Jacqui the feeling is mutual. She tells me, "We've just listed these bats as endangered because of white-nose syndrome, and now we're giving out takings permits, and the environmental community is not coming up to the table to fight that? Or to at least have a say?" She's referring to state wildlife regulators who gave wind developers permission to kill protected bats—permission usually reserved for scientific or educational purposes. "Where's the rest of the environmental community on all of this?" Jacqui asks.

The rift that has emerged between environmentalists seems to follow a fissure between the local and the national or global, and the rift relies on different politics of space and place.[25] For nationally facing public figures, it is impractical to defend local mountains in the face of global climate catastrophes. It seems hypocritical to assert that wind turbines are just fine for big flat states but poorly matched for small mountainous states. But for state and local environmentalists in Northern New England, mountains are tried-and-true symbols around which to mobilize the public. They should not become casualties of a dubiously effective one-size-fits-all technology simply because it seems selfish to defend them.

Wind opponents' experience as spectators to development in their own backyards reflects a common pattern in renewable energy decision-making. One review of the academic literature notes that public opposition to renewable energy has become so prevalent that calling for more public involvement may seem to be counterintuitive to reaching renewable energy goals.[26] As the authors note, however, this sentiment misses the point entirely. Citing a 2011 report titled "Democratizing the Electricity System," they note that many wind opponents are in fact staking a claim against the centralized electricity system that exports both the electrons and the profit to distant locales without really buoying local economies. The report states "local ownership becomes the key to unlocking local support."[27]

The preference for local ownership is hardly unique. Neither are the preferences for transparent policy and regulation and for accessible democracy. How has the renewable energy industry so effectively shifted these issues to the background? In part, the explanation is that the issue has been reduced to a matter of local aesthetics versus the global climate imperative, which was exceptionally easy to achieve given Northern New England's history of exceptionalism and sensitivity to outside manipulation. Places that literally sell themselves to tourists and seasonal residents cultivate a sense of local purity that any development is perceived to disrupt. But mountain defenders are especially easy to criticize in the global context. They are White. They are relatively privileged. They are the most protected from sea-level rise, the climate change event that will impact the most people in the shortest time.

Conclusion

Romaine Tenney, the farmer I learned about in fourth grade, the one who took his own life and set his farmhouse ablaze rather than see it bulldozed, has been on a lot of people's minds around my hometown, more than fifty years after his death. In October 2019, Ben Fuller, a local musician now based in Nashville, released a song about Tenney called "Spark." The comments on Ben's social media posts show that the chilling story still strikes a chord among many locals, probably because we see a little Romaine Tenney in ourselves, in how we love our home state. We spent childhood summers exploring in the woods and winters sledding deathly steep hills. We learned to drive over washboards and through deep muddy ruts in what passes for springtime. We know the smells of a leaf pile in autumn, and the indescribably perfect steamy-sweet air of a sugarhouse. We understand how a farmer could love his land more than anything and how losing it would be a death sentence. We drive our cars over the steamrolled remnants of Romaine Tenney's house when we access Interstate 91, which means that those of us who remember his story have struggled to reconcile "progress" with our own deep connections to home.

Social media has given new voice to these internal negotiations, strengthening their social importance by fostering collective mistrust of outsiders and authority. The last remaining physical piece of the Tenney farmstead is a hulking sugar maple rotting at the corner of the Exit 8 park-and-ride off Interstate 91, where carpoolers meet on their way to work. Vermont's Department of Transportation ordered the tree's removal in the interest of safety, but on the heels of Ben Fuller's new song, the order met with immediate disapproval. Someone erected a small white cross at the base of the tree, behind traffic cones and yellow caution tape. On Facebook someone vented that if the tree could not be saved, it should be left alone to meet its fate naturally, where it grew from seed—the fate that progress robbed so cruelly from Romaine Tenney.

Rural New England is full of communities routinely held captive to a map of progress plotted in faraway cities. Bulldozed homesteads under the interstates, flooded homes under reservoirs, cellar holes in national forests, leaf peepers and skiers clogging the roads, and tourist industry wages are all a legacy of regional land use plans. It is, therefore, impossible to understand the region's tensions around renewable energy without fully appreciating the way that development politics live inside the collective memory of rural communities. But digging through history to map rural communities' feelings of loss and injustice begs the question, how far back in time and how far afield in space should we look?

While it is tempting to embrace the mythology of the "Vermont way" and anoint ourselves the protectors of a landscape we care deeply about, there are deeper truths about the landscape than what we see today. Millennia-old knowledge of the region's natural history was lost with the displacement, genocide, and silencing of Native Americans, who by most accounts were better stewards of the landscape than European settlers. It was, after all, colonists' unchecked deforestation and intensive agriculture that spurred the federal government to create New England's national forests in the first place. Perhaps the region would not be relegated to "vacationland" if its residents had not destroyed what was there before colonists arrived—the enormous trees, the bountiful fisheries, and the peoples who had mastered the sustenance and myriad uses of these resources over millennia.[28]

In light of these historical facts, how rural Americans lay claim to landscape needs tempering, because even in pristine places, European settlers and their descendants have failed on multiple ecological and social counts.

Nowadays, most Northern New Englanders do not realize that by using public transportation and shared heating and cooling, the average resident of Manhattan is having a much lighter impact on climate change than the average local. Indignant environmentalists in rural communities need to focus their rage on energy planners who enable car dependence and petroleum heating, not people who live in cities. In other words, charting a just energy landscape will be far more complex than simply ceding power to rural, white communities if they subscribe to the same benchmarks of progress and merely displace their social and ecological impacts elsewhere. The world is too small and interconnected to think that our energy is pure and just simply because we can see where the electrons are being added to the grid. Woven into Northern New England's energy rhetoric are discussions about *regional* fairness and equity that we have not sufficiently processed in the public or in social science research. Northern New England's rural residents have an identity that in many ways was formed in opposition to nearby cities, which shines light on larger questions for society to consider when designing its energy future. Should rural places be required to disproportionately host renewable energy infrastructure? What, if any, is a fair amount of compensation? Conversely, is it fair that by resisting energy infrastructure, rural people exert disproportionate influence on regional and national energy policy?

In the chapters to come, I use the example of wind turbine resistance to argue that talk of a "just" energy transition should probably not privilege local and regional conversations, which tend to prioritize finances and distract from global inequities. Our energy status quo and its questionable morals are of course the products of a complicated past, but as the world has shrunk and climate science has evolved, we have a better grasp on the consequences of a centralized electricity model that prioritizes low-cost electrons. The argument I build toward is that a just energy transition cannot simply materialize with the advent of new technologies. Technologies are social artifacts created by actors and embodying their interests. For energy technologies to sow justice they must embody social aspirations with justice at the core. Using the case of ridgeline wind, I argue that mitigating climate change is a project that demands greater unity across distant places and diverse peoples than what the North American energy transition has inspired so far.

For North American consumers, this point of view means questioning not just where our electrons come from but reckoning with why we feel entitled to use so many of them, so cheaply. These simple questions point back to the structural origins of our climate crisis. Our relationships to energy are not natural artifacts of our survival, like the need to heat our homes. The ways we think about and utilize electricity are born from a century-long marketing campaign to encourage buying more of it, not because we *need* it, but because every watt puts another few pennies in someone's pocket. Sometimes that pocket belongs to a shareholder, sometimes an executive, sometimes a political appointee or the elected director of a co-op. Thus, the problem is not necessarily that electricity is capitalist; the problem is that our electricity systems have been designed to facilitate cheap and transportable electrons and are the foundation for all other economic activity, in effect centralizing political and corporate power within electrical power. Even amid talk of "transition," our energy systems remain the same public-private administrative knots that have groomed us to believe that electricity should be cheap and plentiful.

For those who fear shortage or scarcity, asking for prices that reflect electricity's true costs is probably more damaging to the businesses built around cheap electricity than to our lifestyles. Centralized electricity is not really a necessity. It is the tasks that we ask electricity to perform that have become necessary for our modern lives. It is hard to defend the Western world's gluttonous consumption habits built on paltry wages for the global poor and catastrophic pollution, but it is easy to envision a future in which our energy consumption can be drastically reduced without having to sacrifice much in the way our lifestyles and livelihoods. The real challenge is resisting the scare tactics deployed to prevent us from trying.

4

But What If . . . ? Wind and the Discourse of Risk

• • • • • • • • • • • • • • • • • • • •

Close your eyes, and picture a wind turbine on fire. Shrouded in thick black smoke billowing four hundred feet above the ground, the blades are backlit in an orange halo of flames as burning oil drips to the ground. Images like this occasionally appear on an antiwind Facebook group that I follow in Northern New England. Although the region's turbines have not experienced fires, explosions, lightning strikes, or collapse, photos of those events from elsewhere are welcome fodder for the group's members. One contributor shared a photo of a burning turbine, with the question: "What was the line from Iberdrola about wind turbine fires? 'That's what insurance is for,' if I recall?" A second poster replied, "Yes that is what I was told by a rep from Iberdrola at the meeting in Danbury NH 'That's what you have fire insurance for.'" A third poster quipped, "Whose insurance?"

In the pasta fight of wind opposition, wind opponents sometimes cite the threat of fire. They point to the hundreds of gallons of motor oil in each turbine and the potential for climate change to make Northern New England's forests dryer and more vulnerable. Wind developers usually assuage these concerns, but the flippant response about insurance highlights a widening gap in the logics used to manage risk and uncertainty. Wind developers take rationalist measures to mitigate risks. They pay to train local fire crews and buy them necessary equipment. They also pay insurance premiums to protect their property and stem their liability in the event that a turbine catches fire.

Many wind neighbors are unconvinced and uncomforted by these rational calculations and skeptical of the science behind them. Understanding science is of course crucial, because *potential* fires cannot be predicted without knowledge of engineering and statistics on past fires, which means that data is equally crucial and that wind developers hurt their cause by not providing it. Turbine fire statistics are so guarded that a company selling fire suppression systems to the wind industry published a report pleading for more transparency. The report cites widely ranging estimates of fire frequency, from as few as one fire in every 2,000 turbines, to one fire in every 15,000 turbines.[1] The result of this ambiguity is vocal public conflict laden with uncertainty.

Uncertainty, however, is a tool utilized craftily in wind debates. Uncertainty was central to the region's wind development in the first place. Promoters justified wind energy primarily by citing long-term uncertainty about electricity price and supply. Wind opponents refuted this logic with their own calculations, spinning off more uncertainty. They hired their own experts and questioned technical elements of wind turbines' operation. Like developers, wind resisters also confidently reference worst-case scenarios, like skyrocketing prices, as a smoke screen to enclose the veracity of their claims and shut down their opponents' credibility.

In this "fake news" era, it is hard to know whom or what to trust in these debates. The public is left to make sense of changing energy systems in cultural spaces flooded with information that is often detached from its political and financial origins. Karen told me over coffee, "The trouble is that some big corporation with good PR people and a team of paid wildlife biologists and analysts and so forth can make a case for their project being good, but it could be false."

The brokering of truth happening in Northern New England's wind debate exemplifies a social theory known broadly as "risk society," which holds that Western society has reoriented around fear and distrust in response to looming scientific and industrial innovations.[2] This chapter invokes risk society to make sense of simultaneous attempts to persuade voters, officials, and electricity ratepayers that wind energy is the salvation and the scourge of our futures. The chapter focuses on public debates about wind turbine costs and carbon accounting, which are key to understanding the region's wind resistance overall. The scope of this uncertainty should, if nothing else, drive home the point that the foundations of knowledge underlying key claims need to be examined thoroughly when we make our energy decisions.

Outlining Risk Society

Renewable energy, as politicians and news media present it to the public, is a strategy to mitigate risks associated with fossil fuel depletion and climate change. While planning for crises is prudent, risk society theory advanced by German sociologist Ulrich Beck observes that modern society is becoming ever more structured around measuring, managing, and mitigating harmful events that have not yet transpired, resulting in rampant fear and mistrust.[3]

Modern industrial society, Beck argued, must grapple with by-products of economic progress that threaten its very survival. An outgrowth of industrial society, Beck's risk society is distinguished by a preoccupation with threats that supersedes concern with matters of equity and justice. In this state of preoccupation, the institutions of risk society launch rationalist plans to study, quantify, and subdue the uncertainty of threats through risk management and planning. And yet despite these practices, members of risk society fear the unknown, express skepticism about science, and feel a sense of abandonment and mistrust for the government that fails to protect them. For example, think about how a wind developer blithely citing an insurance policy premised on the rarity of turbine fire offers little comfort to the people who live near wind turbines and see images in their news feeds of every turbine fire on the planet. Energy policies that put people at risk are seen as irresponsible and the government that promotes these policies as untrustworthy.

Risk in Everyday Lives

The insurance industry offers compelling evidence of risk society. Countless aspects of modern life use insurance policies to hedge against uncertainty. Insurable assets now include body parts, pets' health, crops, subprime mortgage-backed securities, sports betting, and human behavior and relationships. To mitigate insurance costs and appease underwriters, institutions now form offices of risk management that reshape their normal operations around unknown events. With an army of risk-reduction experts sniffing out potential hazards, iconic artifacts have disappeared from our lives. Diving boards, winter sledding hills, deep swimming pools, and iconic playground equipment are just a few examples.

As Beck understood, the threat of climate change constitutes a central orientation in the risk society, as institutions transform to plan for and accommodate the uncertainties of an altered climate. Again, nothing plants this theory in reality better than the insurance industry and specifically the reinsurers who underwrite insurance retailers. Climate change, regardless of one's beliefs about its existence, has already raised our insurance premiums because reinsurers have been tracking sea level rise and incorporating climate change predictions into their pricing for years. One can feign disbelief and denial, as many do, but belying that rhetoric, insurance customers acknowledge climate change through their climbing premiums.

Beck's work also explores how uncertainty changes the way individuals think and behave. While modern institutions prepare for risk by executing tactical plans, risk is manifest in individuals as fear, which devolves to confusion and mistrust. Even if we choose to ignore climate change, we receive myriad signals to maintain an elevated sense of concern with modern life. Exposés on television news present stories about product failures; opaque and threatening technologies; and exposure to toxins in air, water, and personal products—generating imagery of modern life in which there are no safe things to buy, use, or eat. This tendency is reflected in virtually every type of marketing to every type of consumer. Film and television embrace terror, crime, and postapocalyptic plotlines. Reality TV depicts survivalists and "doomsday preppers" on multiple series and networks. Survivalist-themed resorts exist. Costco sells emergency food rations in bulk alongside organic olive oil and tennis shoes. All of these beacons of uncertainty matter for renewable energy because they drown out threats associated with global climate change and ecological degradation, making them appear as abstract possibilities in a distant tidal wave of other possibilities.

Risk and Science

In chorus with media voices teaching us what to fear and how to protect ourselves, we hear the prescriptions of modern science, allaying concern with relative calm and the aura of value-free objectivity. Understanding why wind turbines have become so contested requires understanding how the public has been taught to make sense of science and scientific experts in the broader culture of fear.

Science has become so complex that trusting in it is more important than ever, precisely at the same time that trust is waning. Science education is grossly unequal. Science funding is increasingly tied to commerce. Training in basic science is less and less sufficient for comprehending the technologies we rely on, helping to harden a divide between tech users and tech creators. Science and scientists remain so foreign to most Americans that one study found fewer than 20 percent of Americans have ever met a scientist personally, and fewer than half of Americans can name a living scientist.[4] Not only is the public poorly trained in understanding and interpreting scientific research, but science is also increasingly specialized and technologized, so that even practicing scientists are seeing their own areas of expertise become more narrowly focused and more distant from others'. More and more, we have to take people's word because we are excluded from the specialized knowledge on which technologies and medical treatments are premised.

Being distant from science strengthens the role that fear of the unknown plays in our everyday lives, because we encounter science's limitations differently from its possibilities. Far from being intrinsically objective, scientists have agendas coded into their disciplines, research paradigms, and methodological approaches. Even scientists with flawless integrity must conform their research proposals, questions, and findings to conventions shaped by the politics of science funding. The public is poorly equipped to make sense of the political culture of science, but through courtroom dramas on television, the public sees that both sides of a legal battle can locate experts for hire to substantiate their case. Lacking intimate knowledge of scientific terminology and methods, many citizens have become suspicious of any expert with strong convictions, tragically believing all biases are equal. Doubting claims that evoke fear is also a useful defense mechanism. As a threat grows in its scope and severity, we cling to a hope, no matter how irrational, that the experts are wrong. It is in this milieu that a vocal minority of Americans question the legitimacy of climate change consensus.

Shouldn't affluent democracies protect citizens from risks like bad doctors, unproven medical treatments, and damaging forms of energy? Most of us know that regulatory and oversight bodies exist, but the media exposes us to countless examples of their failures. Beck theorized that states reinforce risk society when they fail to protect citizens' health and safety or hold those who endanger us accountable.

Despite extensive bureaucratic apparatus to measure, regulate, and mitigate risks, the state is often rendered powerless. For example, the U.S. Environmental Protection Agency oversees the release and storage of toxic substances, but Americans are still sickened, injured, and killed from exposure to harmful substances in catastrophic numbers. Because the state has opted not to follow the precautionary principle, stipulating that practices be proven safe before they are permitted, the onus falls on individuals to prove something is unsafe. This requires a formal legal complaint and expert testimony to attest to the scientific mechanisms of harm, while the perpetrator of harm carries no reciprocal obligation to prove products or practices are safe. The result is a mistrust in the capacity of our governments to enforce laws that keep us safe and secure from harm, which Beck sees as a strong incentive for citizens to eventually demand reforms.

How enveloped in risk society are we? Americans' perceptions of medicine offer a compelling example. In an early draft of this book, I described several ways that modern Americans confront contradictory messages from doctors and health experts, but I can replace all that with a reference to COVID-19 and the issue of masks. In less than three months, we bore witness to whiplash-inducing flip-flops on the issue of face protection. Citing no evidence from research, in the early stages of the pandemic countless physicians used various platforms to advise that masks make the public overconfident and more likely to pass the virus from their hands to their face. Infectious disease experts from across the world cringed, particularly those familiar with strong population-wide evidence from Asian countries that masks are indeed effective at slowing airborne viral transmission. Eventually, most of the world caught on. In hindsight, many claim their early warnings were merely meant to protect the mask supply for health professionals, but given that even fabric scraps make adequate masks, it seems likely they were also ignorant of an area of medical expertise far outside their own.

Mistrust in medicine is deeply entrenched and rampant. We give doctors godlike status but we also second-guess them. Our specialists contradict our primary care doctors. Google contradicts them both. The pandemic has made this problem much worse. We are watching doctors learn on the fly. We hear constant disagreement over testing rates and treatment protocols. Viral modeling is widely cited but is ever-changing as new data are added. Experts rarely contextualize these fluid models to educate the public and our leaders about effective strategies to reduce transmission. Instead, people are shown a graph with a scary number but are rarely told that their actions are what decides how scary the number is. In the media, the virus is discussed as an invader colonizing our bodies, which makes people feel powerless. Many Americans have chosen not to wear masks or avoid crowded indoor spaces because so many experts and authority figures have opined in opposition to one another about how deadly or how contagious the virus is or how wearing masks affects people. Much of this "expert" advice arrives through social media memes of unknown origin, but few people would know how to make sense of the different origins anyway, because all science seems somehow biased or corrupted.

Our governments are supposed to protect us, but even that doesn't seem to be happening uniformly around the world with COVID-19. The U.S. federal government has shifted in and out of willful denial, interrupted by a brief window of sensible precaution. In the process, we've seen the reputations of the Centers for Disease Control and Prevention (CDC) and the World Health Organization (WHO), once scions of public trust, dragged through the political mud. Every day we see pictures of Donald Trump without a mask, but we do not see the army of nurses testing him and everyone around him daily. The message sent to the public is not that the Trump government is inept, which it has shown itself to be; instead, the message is that COVID-19 is not a threat that we need protecting from. The death toll and transmission rates, especially when compared with other rich countries, tell a very different story, and so does Trump's daily testing for he and his staff

This terrain of fractured knowledge, knowledge sources, and perceived biases takes root in fragmentation across fields of science, which creates multiple coexisting truths that can be marshalled and weaponized for political and financial gain. This fragmentation helps explain the anxiety we feel as we look for solid ground on which to plan for our health and make sense of our fast-changing world. This fractured terrain is essential context for understanding why wind energy has been resisted in Northern New England and why that resistance led to eroded trust in governance.

Beck on Climate Change: Risk and Wind

Northern New England's wind opponents contest with great sophistication what they see as the manipulation of uncertainty by "Big Wind" in order to promote projects without clear advantages in terms of carbon emissions or costs. A public dialogue has emerged between several vocal and active camps, each making claims about rationality and scientific objectivity. In what follows I describe the discourse around risk and uncertainty as reflected in the domains of energy pricing and climate change.

Hedging: Energy Pricing

The case for renewable energy in Northern New England was motivated by volatile electricity prices. Compared with decades past, current fossil fuel prices are volatile because of complexities in extraction processes, tumultuous geopolitics, environmental laws, and rampant speculation. In a nutshell, electricity has become much more complicated. Any business or home can sell electricity to the grid, requiring grid managers to use complex forecasting to predict supply and demand. Forecasts account for short-term summer and winter weather, long-term cyclical climate shifts, foreseeable changes in technology, global political uncertainty, and now progress toward efficiency.

Wind turbine development proliferated after 2008 as a result of formal policy interventions that reformulated the risk calculations made by energy companies. Construction subsidies, tax credits that convey to future owners, and long-term purchase contracts built a scaffold of stability into the wind energy domain. In spite of wind's intermittency and the general unpredictability of weather, the U.S. government's incentive programs turned wind into an energy source that seemed predictable in comparison with the tumultuous fossil fuel markets.

Discussing Vermont's renewable energy law with a former member of the state legislature, I was given a window into the bill-making process through which lawmakers and wind industry interests were able to agree on rate-setting and tax-stabilizing measures that accommodated the corporate aversion to risk. According to this account, electric utilities had been unwilling to move on renewable energy without feeling certain that costs would remain stable:

> We said, "50 megawatts of power could qualify for a fixed price." Long-term contracts set according to the price of their own technology. Suddenly we had developers who could go to a bank and say, "I'm getting paid this, for this long." They were able to finally fill in the line items that could make it possible to

evaluate a business plan and make some money, and get rid of that uncertainty about whether the contract price would drop below what they could make sense for what they had to capitalize. Then another uncertainty, because I was always listening to, frankly, renewable developers, what was in their way. Another uncertainty was tax. How would they be taxed. There was no uniform way. No one knew how to appraise the value of these systems. . . . People were appraising these things in wildly different ways, so you didn't know what your tax would be which is, of course, another line item on the spreadsheet.

Wind opponents see these negotiations differently. Opponents poke holes in the economic narrative used to promote and justify new installations by identifying the uncertainties in developers' strategies to mitigate uncertainty. Foremost, opponents question the estimates used to predict a site's energy potential and the methods used to assign capacity factors to new wind turbines. National maps of wind potential show that only narrow slivers of ridgeline in Northern New England receive winds of sufficient strength for commercial energy generation, leading many to believe that developers in the region overestimate the productivity of their sites in order to win approval and long-term contracts.

Moreover, opponents note that wind developers fail to include a number of other relevant parameters into their cost projections. In Vermont, the infrastructure required to tie wind turbines into the grid has not always been accounted for. In the Kingdom Community Wind project, synching twenty-one remote turbines with the electrical grid was precarious because of volatile summertime demand and inadequate power lines. Without a $10-million synchronous condenser, the new turbines were at risk of spinning idly on even the windiest of high-demand days because the regional grid operator would opt for more stable sources of power. Only after this scenario became a reality on one scorching summer day in 2012 did the project's managers purchase the condenser. This expenditure, framed as an afterthought by opponents, is cited as additional evidence that wind developers overstated the economic case for wind. With conspicuous cost overruns, consistently overstated capacity factors, and the potentially under-estimated costs of eventual decommissioning, wind opponents character-ize the pursuit of wind energy as a *source* of economic uncertainty rather than a solution for it.

Renewable Energy Credits are an additional source of price instability because of the extent that renewable energy rates now depend on RECs' future value, which will vary if the past is any indication. A REC's worth is affected not simply by supply and demand within individual markets but also by the fine print in energy policies that underlie demand. For example, in lieu of RECs, some states like Vermont accept "Alternative Compliance Payments," allowing utilities to pay a fee instead of actually meeting the goals set by renewable portfolio standards. Alternatively, New Hampshire allows utilities to "acquire" a portion of RECs unclaimed by their owners, meaning that solar owners who fail to specify what happens to their RECs lose them to utilities. Both scenarios introduce uncertainty to the future cost of RECs. Because Vermont's ridgeline projects were financed based on estimates of future REC sales, declines in REC values will yield lower returns to wind developers, which can be recovered simply by charging consumers higher rates.

Northern New England is home to a number of additional stakeholders posturing in discussions of future energy prices, resulting in a chaotic and confusing narrative. The Regional Transmission Organization (RTO) that manages New England's electricity supply (ISO-New England), state grid managers, public and private energy companies, politicians, and energy investment banks must all hedge against uncertainties to protect their particular interests. ISO-NE points to climate change and the potential for blackouts in advocating expanded fossil fuel capacity. Wind developers of course point to volatility in energy prices to justify wind projects. Transmission companies temper developers' promises of low electricity prices by identifying costly improvements needed to accommodate more renewables on the grid.

Wind opponents knowingly create price volatility through litigation, to amplify uncertainty. The following comments, from a posting on the website maintained by a group called Friends of Maine's Mountains, are in response to a developer's decision to give in to a community's antiwind litigation: "Our strategy is no secret, as demonstrated by the people of Molunkus. We want the shareholders of the huge wind corporations to know about the enormously expensive regulatory and legal hurdles they face as they try to rob Maine of its famous wild areas." As declared proudly, by engaging in the regulatory process, testifying at meetings, printing flyers, and making signs, wind opponents have inserted themselves in the calculations of risk made by prospective wind companies. Not only do prospective wind developers have to estimate the costs of construction and production amid the vagaries of terrain, weather, and politics, but developers also must account for the costs of litigation in the face of local opposition.

Wind opponents also harbor suspicion that developers are escrowing inadequate funds to properly dismantle, remove, and remediate ridgeline wind projects when they cease operating. In public meetings and editorials, New England's wind opponents point to shady industry accounting that ignores the ballooning expense of decommissioning wind turbines. Opponents' words stoke the fears of many by painting a dystopian portrait of rusting turbines sitting lifeless on the region's mountaintops, lights blinking at nighttime, long forgotten as sources of energy or aesthetic grace. Articles shared on Facebook groups are starting to draw greater local attention to the fact that most turbine blades cannot easily be recycled and few landfills accept them. Similar worst-case scenarios have now emerged in debates about solar installations, including the fear that shattered glass panels could render land permanently unusable for agriculture.[5]

Beck argued that the risks and rewards of orienting toward risks are not evenly distributed under capitalism. Many electricity companies still enjoy monopoly status in their local markets. All sell their product per unit and have a natural inclination to worry little about price volatility, because they can pass increases on to consumers. Under these conditions, it seems logical to dig deeper when companies hold up future price volatility as a threat to plan against. Using the threat of future price increases as leverage to elicit support for a particular type of electricity generation would seem to be a strategy employed only when that type of generation is lucrative for the company.

Mountaintops, Carbon, and Climate Change

Northern New England's wind opponents address cost because wind developers used cost as a selling point for ridgeline projects. Opponents also focus on what society does not yet know about ridgeline wind's impacts: what we cannot see, what we are not allowed to see, and how those unobserved threats should be assessed and evaluated by developers and regulators.

In progressive enclaves like Northern New England, much of the public appreciates the urgency of climate change, even if they are relatively protected from its effects. Thus, many associate wind turbines with "doing their part" to minimize risk and not exacerbate it. One commenter at the solicitation of public input about Vermont's 2015 energy plan exemplifies this sentiment:

I think the plan is really exciting. There are so many things going on in Vermont. It's really exciting to look at all of that and see how much quality work is happening. As I looked through the plan, I kept feeling like it needs a stronger sense of urgency. Climate change has been increasing more rapidly in recent years, the effects of climate change, the storms, the fires, and what's in the pipeline for climate change and it's going to be coming at us even faster, so there really isn't time to gradually work into these reductions. Every opportunity should be taken to strengthen the plan.

This sense of urgency has become especially pronounced since Tropical Storm Irene eroded Vermont valleys and pummeled our storied covered bridges in August 2011. In the wake of Irene, Vermont officials used the event to promote renewable energy. Hurricanes are central to societal orientation toward risk because climate models suggest that warming is altering ocean temperatures and currents, triggering storms of greater frequency and intensity. While some years bring few if any storms, models predicting larger, stronger hurricanes have proved accurate. Accordingly, after Irene, discussions about energy in Vermont were laced with references to the storm. Several Vermont businesses impacted by the storm testified before legislative committees in support of a carbon tax, including a popular brewery that faced high cleanup costs and now grapples with uncertain costs from climate-impacted hops and barley crops.[6]

While it seems obvious that wind energy is an important part of mitigating climate change, wind opponents have not uncovered much evidence to support that assertion in Vermont. Whereas a carbon tax would directly promote in-state reduction of emissions, Vermont's wind policy has played fast and loose with greenhouse gas benchmarks and goals, leaving the policy vulnerable to criticism. Online commenters begrudged Vermont's former governor Peter Shumlin for using Irene as a promotional tool for new wind projects. Said one, "And then there's the 'save the planet' crowd who seem to have been brainwashed into thinking that throwing lots of solar and wind up on the big grid is somehow going to prevent another Irene. It's more a religious vote than anything that will be meaningful in terms of addressing climate change."

As this comment illustrates, some take issue with using Tropical Storm Irene to sell ridgeline turbines as investments in climate risk reduction. It could be seen as "religious" because references to Irene tap into the strong sense of unity that people felt when helping their neighbors recover from the storm. Using unity to sell renewable energy probably strikes most environmentalists as brilliant messaging, but for wind opponents this is a lie propped up by selective accounting. Ralph tells me that he is fed up with politicians telling the public that "if you begin reducing carbon emissions now, in Vermont, it will protect us from Irenes in the future. It will put us back on the kind of winters that we used to have in the 1970s. . . . That's total bullshit, and we know it." For Ralph, this is pandering to people's longing for winters from their childhood while keeping them in the dark about the hard reality of climate science—the reality that normalizing the climate from our past carbon emissions could take thousands of years.

On the subjects of wind, uncertainty, and the weather, it merits noting that some wind opponents are now wading through scientific analyses on wind turbines' impact on weather patterns. Posters to one Facebook group poked fun at a Texas congressman who proclaimed that the wind is finite and that turbines slow it down, exacerbating warming. Subsequent posters, however, cited studies by the National Aeronautics and Space Administration (NASA) and Harvard University finding that wind turbines do interfere with nighttime inversion patterns that cool the earth's surface, resulting in slightly warmer temperatures on the ground beneath wind installations.[7] Several studies confirm this local effect. In research that used computer models to study the effects of building wind and solar installations in the Sahara Desert, scientists found evidence that with enough wind and solar, sufficient precipitation would fall to green the desert with carbon-sequestering plant life, minimizing the impact of slight elevations of temperature.[8] What is essential to note, of course, is that mild increases in ground temperatures are redistributions of existing temperature differentials in the atmosphere, not systemic increases to the atmosphere's retention of solar radiation, such as with greenhouse gas emissions. The way that wind energy's meteorological impacts are packaged for public consumption illustrates risk society's tendency toward confusion and mistrust. While some wind opponents in liberal New England label the Texas congressman an "idiot" for his simplistic grasp of science, others engage with the nuances of his argument to reveal the source of legitimate confusion emerging from climate experts.

Wind opponents use a vocabulary of uncertainty and risk when contesting the premise that ridgeline turbines are carbon-neutral or climate friendly. Rather than free the electricity grid from dependence on fossil fuels, wind turbines in Northern New England need counterbalance by natural gas power plants that can be brought on- and offline to accommodate the wind's intermittency. Christine recounted participating on a panel in which fossil fuel backup became a sticking point that the panels' wind developer had difficulty addressing:

> So we get to the Q and A, and one of the people in the audience stands up and says, well, the opposition side made a big deal about how these aren't addressing carbon reduction, and you guys didn't rebut that. What do you have to say? [The wind developer] took the question. And first he goes off on a tangent . . . "When you make a wind turbine, it uses less carbon" than I guess "to build a natural gas power plant." That was his point . . . then he gets on to the grid: "Yeah, it's stupid. We need a smarter grid." He agrees with me! We need peaking generators. This whole fleet that we have, nuclear, and you know . . . it's got to be replaced by peaking generators. You have to somehow create incentives to companies to want to build. They're very expensive!

Peaking generators are electricity plants, typically burning natural gas, that can be powered on quickly to supplement baseload power plants when demand on the grid spikes. They are essential to efficiently running a regional grid with lots of intermittent power sources like wind and solar. Without sufficient peaking capacity, renewable energy can be curtailed, or denied from entering the grid, to mitigate volatility. Without efficient, fast–ramping peaking generators, it can be difficult to minimize greenhouse gas emissions, because the power plants standing by to take over for the wind may be burning fossil fuels inefficiently.

A filing by North Carolina utility Duke Energy claims to confirm this.[9] The company reported that powering its plants on and off to accommodate solar electricity increased certain emissions because the plants must pass through high-emission phases of operation. Duke Energy reported that nitrogen oxide (NOx) emitted from its plants with solar on the grid exceeded amounts released under purely fossil fuel generation, and the utility speculated that adding more solar to the grid risked tipping the balance on carbon dioxide (CO_2) as well. After its report garnered negative attention, Duke Energy released subsequent statements clarifying that it was in no way shirking its commitment to renewables. While an electric utility may not be the most trustworthy source for these calculations, there are no other sources for the necessary data. If Duke's predictions are even slightly accurate, they lend credence to wind opponents' suspicions that turbines are much less effective for climate mitigation than we think. This uncertainty makes carbon accounting for renewable energy a contested and confusing exercise.

Another source of confusion has come from electricity output estimates that wind developers promote before they build but never follow up on. Wind opponents contend that many turbines have generated less electricity than originally promised, hinting that developers embellish turbine capacity from the start. A Harvard study estimates that industrial wind's mean production is less than 33 percent of the maximum "nameplate" capacity possible under nonstop production, which is lower than capacity factor calculations made by the U.S. Energy Information Administration and lower than industry estimates. The same study's analysis of 411 onshore American wind installations estimates that industrial wind's power density—its electricity output per unit of space—was lower than many previous estimates, which varied by a factor of seventy. Wind turbines were found to have a power density roughly one-tenth that of solar installations and even less under marginal wind conditions.[10] No one is sure of the disparity between the world's estimated wind capacities from its observed performance, much less how much carbon is actually displaced from the electricity grid.

Power density is particularly important when the space occupied by wind turbines has been converted from another climate-friendly use. Wind opponents question the carbon accounting used to justify putting wind turbines in mountain forests, as opposed to pastures and prairies. Ridgeline wind in Northern New England requires blasting several vertical feet from granite peaks to level them as well as blasting and leveling across mountain slopes to build the access road. (Christine estimated that nine thousand gallons of fuel were injected in granite for blasting at one ridgeline project.) Once in place, turbines are implanted into several tons of concrete, which is the second-highest CO_2 emitting building material after aluminum. And of course, what becomes road and flattened ridge was once thick, soil-nurturing, dense temperate forest, which sequesters carbon naturally. These carbon-intensive activities come in addition to the energy required to fabricate, transport, and operate the enormous turbines, and in comparison with places with stronger, more consistent wind and more accessible terrain, these activities dramatically alter the accounting of wind energy's ecological value.

These observations have motivated Dr. Ben Luce, a Vermont physics professor who once directed renewable energy at Los Alamos National

Laboratory, to speak out against ridgeline wind in several public appearances. On one radio call-in show, he proffered:

> I think that a lot of well-meaning people jumped on the bandwagon for utility-scale wind in Vermont, very quickly, and the industry really took advantage of that and really lobbied our legislators to push through a lot of policies. They built up a policy structure that really fast tracks this technology, but with very little quantitative analysis of whether this really made sense . . . very little really serious thought or concern for the true environmental impacts and really no rigorous and objective comparative analysis of how this particular technology stacked up and compared to other types of renewables technology. It became a kind of a free-for-all.[11]

According to another of Professor Luce's presentations, the region's wind potential is not just hampered by the carbon costs of site preparation or marginal wind strength. Professor Luce also raises doubts about wind turbines' durability in a variety of wind conditions. He references studies showing that their gearboxes are susceptible to damage under turbulent and fluctuating winds, translating into shortened life spans for the turbine and drastically overstated generation capacity.[12]

Poring over the public debate about wind turbines, it is easy to understand why opponents began to doubt that carbon reduction was part of the industry agenda at all. The problem is not simply that greenhouse gas accounting faded to the background. Over time, Vermont leaders appear to have backtracked from any greenhouse gas claims. For example, when the Vermont Department of Public Service faced public criticism for its relatively weak renewable energy policy, representatives made a series of telling statements about the importance of REC trading, which seemed to underscore that any environmental benefits from renewable energy are in fact secondary to the financial benefits. One article reads: "Darren Springer, the deputy director of the Department of Public Service, says the challenge is two-fold: 'what is the most cost effective way Vermont can pursue renewable energy and other policies and programs?' he said. 'And how to do that in a way that maximizes rate-payer benefit, how do we do that in a way that maximizes economic benefit for Vermont.'"[13] Putting the matter in even clearer terms, Vermont's commissioner of the Public Service Department was quoted by an investigative news website as follows: "'I disagree with the characterization that the reason we're doing this is to try and improve global warming,' Chris Recchia, commissioner of the Public Service Department, told *Vermont Watchdog*. 'It is certainly a byproduct of it, and a help, but primarily why we're doing it is to have stable energy pricing and really secure energy resources that are renewable in our state.'"[14]

At the core of many wind resisters' arguments is the simple belief that the urgency of our climate crisis means we should not let wind developers off the hook when it comes to clarifying the ecological impacts of wind projects; but even the wind industry is inconsistent about carbon and climate. For example, in defending against common wind opponent accusations, Patriot Renewables, the developer behind four wind projects in Maine, including Canton's, maintains a "resources" section on the company's website with four editorials. The editorials cover turbines' economic viability, health impacts, subsidies, and the general need for transmission upgrades across rural Maine. None of these four documents mentions carbon emissions, and nowhere on the Patriot Renewables website can one ascertain the four projects' productivity, much less their carbon footprint. The same is true for promotional materials describing most of the region's wind facilities. These materials usually include general statements about wind-generated electrons displacing fossil fuel electrons, but nothing more detailed about specific projects' carbon accounting. Wind facility websites, press releases, state certificates, and even documents that provide detailed production data make no reference to greenhouse gas reductions. Wind companies' websites highlight wind turbines' centrality to the current energy transition portfolio, but rather than justifying this centrality with evidence of climate change mitigation, the companies make the case that turbines are stable, long-term investments that keep electricity rates low. It is plausible that wind companies are attempting not to agitate climate change deniers, but wind opponents contend that carbon accounting is absent either because climate claims cannot be verified or because they are known to be false.

There is scant research investigating the validity of these concerns but strong evidence to suggest that they take root in scientific uncertainties. Other industries have added environmental scientists and corporate sustainability experts to track complex supply chains and manufacturing processes to limit carbon emissions. Why has the wind industry not followed suit? With the emergence of climate change denial has come an apparent gag order on messaging from the wind industry that promotes climate justice. Why is it so difficult for the corporate scions of the green energy movement to move from conversations about climate justice to a transparent accounting of climate justice?

As chapter 7 explores, the answer could be that there is too much variability with wind energy. Energy scholar Benjamin Sovacool and his team were commissioned by a turbine manufacturer to conduct a detailed carbon accounting of wind turbines in Norway and Denmark. They found, much to their funder's disappointment, that the technology yields as much as 1 million Euros in environmental damages, largely carbon emissions, depending on materials and installation.[15] This means, according to the authors' calculations, that as much as five years, perhaps a quarter of turbines' life span, is required for each unit to achieve net carbon reduction. This figure will of course be higher for turbines erected where forests once stood and backed up by petroleum-fired power plants.

The problem is not just wind. Residents of Northern New England have borne witness to considerable debate about hydroelectric dams and greenhouse gas as Maine considers allowing the energy corridor needed to link Quebec and Massachusetts. The issue has been clouded by competing scientific claims about hydropower's climate impacts.[16] While scientists acknowledge that flooding forests creates potentially high greenhouse gas emissions, Hydro-Québec has countered that such emissions are far less in boreal climates than in the tropics, and in some cases the company has argued its emissions are negligible in contributing to global climate change.

Weighing in on the debate, an MIT earth scientist and part-time Maine resident published an op-ed in Maine's *Portland Press Herald*, cautioning readers that "to answer this question correctly, we must use the best available science. The *Press Herald* should avoid passing along Hydro-Québec's misinformation. Either the utility officials who claim their power is carbon-free are ignorant of the science published by their colleagues, or they are ignoring this established science in their attempt to sell power."[17] After accounting for the rate at which Quebec's flooded bogs and biomass decompose when submerged under shallow water, the op-ed asserts: "Extrapolating for a century, Quebec's hydro is about half as dirty as gas—something of an improvement, but in no way 'carbon free.'" The scientist recommended that hydropower be used to replace coal power in neighboring New Brunswick to have maximum mitigating effect on climate change, leaving the northeastern United States to pursue cleaner options.

Conclusion

Wind turbines in Northern New England exemplify Beck's risk society framework. Their appearance on the landscape has given neighbors and voters cause for suspicion on a number of technical matters that the average nonspecialist is ill-equipped to interrogate. The financial accounting used to justify putting wind turbines on mountaintops, the science associated with measuring wind turbine sound and minimizing impacts with birds and bats, and the physiological relationship between turbines and health are all black boxes requiring weigh-in from experts, who may ultimately not agree with one another.

Wind supporters and opponents alike debate minute details about turbines' costs and carbon emissions in dozens of online exchanges in newspaper comments sections. Detail-oriented posters seem to have accessed every morsel of available data to calculate their own projections of what electricity should cost and how polluting various sources are. Commenters pick apart one another's calculations, underlying assumptions, data sources, and conclusions. These exchanges exemplify risk society because they reflect a common mistrust of public institutions and frustration with opacity in the energy sphere; and they drip with suspicion of other posters' credentials and biases.

In technology debates, a widening knowledge gap allows discourses of uncertainty to fill in for reasoned scientific consideration. One can recognize how central uncertainty is to different sides' rhetoric by the conspicuous absence of potentially favorable uncertainties from public discussions of energy futures. For example, despite being on the cusp of rapid advancements in battery storage, wind opponents do not acknowledge that the problematic ecological and financial economics of ridgeline wind could soon be rectified. Similarly, few in Vermont have observed that in recent years the state's electricity use has actually decreased as a result of statewide conservation and efficiency efforts. This achievement offers clear evidence that demand-side initiatives can work, and yet few voices in the energy debate look with optimism at the potential to reduce society's demand for its total energy consumption.

Admittedly, it is easy to be cynical given a pervasive shortage of knowledge and data. To make informed decisions about a new technology like industrial-scale ridgeline wind turbines, citizens and elected officials need access to information about the science and scientists behind it. But more than that, they need skills for understanding data and a rudimentary understanding of the scientific method and basic scientific concepts, which are not widely possessed among members of the public or even public officials. Look no further than the members of the U.S. Congress serving on the House Committee on Science who struggled to grasp the simple physics behind sea-level rise when observing that a glass of ice water doesn't overflow when the ice melts. (In response, Jon Stewart, then host of Comedy Central's *The Daily Show*, gently pointed out that our ice caps and melting glaciers currently rest on land.)

New England's wind backlash is different. It is so vexing precisely because so many intelligent, caring people are involved. Not only do those on both sides believe in their own environmentalism, but there also appear to be no differences in educational attainment across the wind energy divide. The problem is, citizens trying to make sense of the issue must drink from a fire hose of information, from innumerable sources claiming expertise but not revealing their personal or financial stakes in the debate. Thus, one's position on the technology likely takes root in competing assessments of future risk and uncertainty that are strategically employed to achieve particular goals in the present.

In this context of complexity and uncertainty, it is not merely society's formal restructuring toward risk that characterizes Beck's risk society but also how individual actors experience and respond to the new structures' mounting complexity. Parts of our lives that were once simple are now the domain of specialists, and in response we succumb to doubt and increasingly reject the once trustworthy authority of science and government out of exasperation and confusion. Beck and his contemporaries, however, were hopeful that scientists could lead the way in acknowledging the boundaries of their expertise and, through this "reflexive modernization," embrace a deliberative process of governing through which we learn about our communities and ourselves.[18]

It can be tempting to respond to cacophonous and contradictory messaging on these matters by relying on our existing political and ideological orientations and beliefs. Wind developers exploit that inclination by promoting their product as a low-cost, low-carbon energy solution for everywhere. Wind opponents are working hard to disabuse belief in wind's purity by throwing at us every argument that will stick. The answer is not to be paralyzed by uncertainty but rather to better map and navigate our new scientific landscapes, accepting that there are multiple forms of expertise and seeking out and listening to diverse experts. Acknowledging our biases for what we find desirable, we have to be comfortable contemplating answers we do not like in order to, at the very least, learn about the origins and assumptions of other perspectives. What is at stake when we let ourselves become overwhelmed with competing claims is tacit acceptance of the energy industry's confident and neatly packaged messaging as truth.

5

Following Power Lines

•••••••••••••••••••••

A Regional Political Economy
of Renewables

Part I: The Money

Not far from my house in Vermont, along the border with Quebec, a farmer
and a wind developer proposed a single four-hundred-foot turbine. The project would supplement the dairy farmer's livelihood at a time when the price
of milk is so low that most dairies had not turned a profit in years. By the
time of its proposal, however, even the small "Dairy Air" project faced opposition. Wind resisters were motivated to stop additional wind development,
having seen mountains blasted and heard the litany of arguments against
ridgeline wind. The project's public meetings, which I viewed online, were
intense, awkward, and a little confusing. One meeting included an exchange
that speaks volumes about the economic complexities of the electricity
market that anger rural citizens. Simply put, communities perceive that
renewable electrons are a product and that unless they assert themselves,
communities will be treated as merely backdrops for making money in the
sale of that product.

In the evening meeting, a member of the public asked: "The power that's generated, where is it going to go? Is it going to stay on the farm? Is it going to stay in Vermont? So many times we generate power and it goes to Connecticut or Massachusetts, or . . ." The question recognized that the market for electricity, even in a rural part of a rural state, is connected through a regional grid to cities. What would rile the crowd that night, however, was not the location where electrons end up, but the exchange of money that would take place as a result, as well as claims to bragging rights over renewability.

One of the developers' representatives attempted to answer the question. Her answer began, halting and uncertain, "the power, is being . . . so, there's two paths . . ." She looked at her colleague, and then, as if it were her first time being tested on this knowledge, she remarked, "I'm going to try this, okay?" She started again:

There's two paths for power, one is the electron, and one is the dollar, right? Who's going to pay for it. So, the state of Vermont . . . there is a contract for the purchase of the power to the Vermont utilities. . . . The electrons, are going to go right out on the wires and on the three phase line, that we will put, that we will upgrade, the existing single pole line, it will be another single pole line but it will have a cross bar, it will look just like the poles that are right outside on the building here, and those electrons will feed into the wires right here in town.

The speaker outlined, with the broadest possible brush strokes, that an ambiguous state entity would manage the accounting of electrons and their prices for Vermont's electric companies. After she offered intricate but unsolicited details about the utility poles, she moved on to a new question, only to be interrupted when someone in the audience called out, "What about the RECs?" Others chimed in: "Wait, wait. . . . But that's part of her question . . ."

Ignoring the crowd's agitation, the wind rep selected another question, remarking, "Well, I'm trying to get to seven o'clock," a reference to the meeting's quickly approaching end. She pointed to another audience member with a raised hand, but that person said, "The last question, about the RECs, do you want to take that?"

Watching this unfold, even an energy novice could tell that a REC is something important. Not only were people in this community, steeped in a regional culture of reticence and reserve, insistent that the REC question be addressed, but the speaker seemed equally insistent on avoiding it. As I will describe at length in this chapter, RECs *are* important. In effect, a REC, or Renewable Energy Certificate, is a coupon for every renewable electron, which equates to green bragging rights. More precisely, a REC is what gives legal validity to the claim that purchased energy is renewable. My informant Ralph referred to Vermont as the West Virginia of renewable energy because RECs are leaving the state and absentee interests are profiting while locals lose their mountains. Later, I will explain how the market for these coupons is the shaky policy mechanism on which our slow energy transition is premised. But first, it is useful to know how the story of the Dairy Air meeting ended, because it illustrates how the public experiences a complex web of economic apparatus in the electricity industry.

With the audience seemingly hell-bent on hearing a response to the REC question, a second facilitator interjected, "It goes with the state contract." Someone responded, "What does that mean?" A third facilitator offered clarification: "The state is purchasing the RECS and the energy." Then the second speaker took back the baton: "The state of Vermont owns . . . the state . . . it used to be called the facilitator . . . VEPP, Vermont Electric Power Producer, VEPP . . . I can't remember . . . it's the state entity that handles these contracts, it takes the RECs."

This is where filmed conversation of the REC issue ends—with a clumsy description of Vermont's energy policy structure but no mention of energy leaving the state, the very essence of the initial question. In fact, VEPP is a nonprofit collaboration between electric utilities and power generators, not a state-owned entity. But the real oversight was REC policy; utilities are only required to buy and sell a certain percentage of renewable capacity, so RECs that exceed the mandate can be sold out of state, banked for a later year, or arbitraged for cheaper RECs. Selling RECs in other states means that Vermont ratepayers may not be consuming renewable energy. In fact, a report by Vermont's Department of Public Service indicates that Vermont utilities sold their customers approximately .1 percent wind power in 2018, despite wind constituting 10.6 percent of energy delivered to the grid that year.[1] A REC, after all, is more than green bragging rights. It is ownership of a green electron, so a REC sold out of state will not benefit Vermont's environmental goals like a REC that is "retired" in state. But these details were never explained.

If all of this is difficult to follow, it is because the electricity industry that has spearheaded the region's wind development makes it difficult to follow. The mistrust that many in Northern New England feel for wind developers is rooted not just in regional history or a love of mountains. Residents' feelings are a response to big energy systems that consolidate political and economic power in ways that defy democratic sensibility. Some wind opponents harbor antipathy toward urban places because the economic relationships connecting them to those places, although murky and shrouded in confusing language, are palpable.

One example of this palpability is industry consolidation. As described in chapter 3, energy conglomerates now own the largest electric utilities in all three Northern New England states, three of which are international. Consolidation in ownership has transpired in lockstep with climbing electricity rates and wildly unpopular corridor projects. And more than simply putting a new logo on ratepayers' bills, consolidation has cost jobs in electric companies, put electricity rates further outside local control, and, as I describe in the second part of this chapter, remapped the landscape of political influence behind wind electricity. It is therefore unsurprising that rural people care about where the energy produced in their backyards will end up.

With REC markets attracting so much attention, it is useful to make sense of the region's energy policy and players as a political economy of renewables. A "political economy" refers to the structures of finance, governance, and authority that facilitate the distribution of capital (money) in a particular place and time. Despite seeming to be freestanding, wind turbines are tethered to many structures. Turbines are tied to the grid in a physical sense, and they are connected through satellites or cellphone towers to control centers monitoring their operation. Economically, wind turbines are equally connected. They are woven into a web of funding and finance, from their siting to their construction and operation. That web helps to explain the legitimacy of questioning why the region's mountains are now generating renewable energy for other states. In the first part of this chapter, I focus on the economic structures underlying wind electricity, and in the second, I focus on politics and political actors.

Neoliberalism and the National Wind Electricity Landscape

> It's a joke, and our [wind developer] sat right there in the public hearings and said, "Frankly, if it wasn't for the tax credits we wouldn't be doing this project"

which tells you right there that it is
just a scam, another one of these green
scams.

Whether you are a wind resister or an energy industry insider, you will probably cite federal subsidies as the linchpin responsible for wind energy's proliferation, as this online commenter observes. More than 90 percent of wind
energy installed in Northern New England as of 2020 commenced after federal incentives expanded in 2009. Because the region's marginally suitable
winds hit only the highest peaks, industrial turbines could not, at the time,
compete with other electricity without incentives. Of course, the story is
more complex, and I will describe later how policy makers use multiple
"carrots" to promote growth of the industry. But it is true that federal subsidies are the largest and perhaps most important carrot, and therefore
subsidies are the most talked about and maligned policy tool that wind
opponents cite when explaining why ridgelines are under threat.

Subsidies are monetary incentives for achieving policy goals, and most
wind opponents believe that climate change is a goal worthy of government
spending. But, it is important to contextualize wind subsidies, and specifically, to assess how they are used, by whom, and in relation to what climate
policy alternatives. Further, I propose that focusing too much on subsidies
opens the door to complaints about renewables' costs, which is usually a tack
that shifts the conversation away from climate. Rather than worry about
what renewables cost, my goal in mentioning subsidies is to shed light on
how wind turbines' costs and benefits are distributed across different actors.

Federal interest in renewable energy as measured by government incentives has waxed and waned over the past fifty years. As wind historian Robert Righter has observed, current levels of federal expenditure on wind
pale in comparison to that seen during the energy crises of the 1970s.[2] Even
with boosts to wind under President Barack Obama's 2009 stimulus plan,
government support for wind was, and is still, eclipsed by fossil fuel subsidies. Relevant to this chapter, however, is the fact that beginning in the
1970s, federal wind subsidies have prioritized large-scale, grid-tied designs.
On the topic of these subsidies, environmental sociologist David Hess, an
environmental policy expert, notes that "the result in the cases of both solar
and wind energy was an incorporation and transformation process that
made the alternative energies compatible with centralized, corporate ownership and transmission via the grid."[3] So in other words, renewables subsidies

have been targeted with an eye toward making large private corporations, manufacturers, developers, and electric companies more profitable when pursuing renewable energy.

Northern New England's experience with wind puts an array of policies on display that favor large companies, and like most economic policy these days, the region's policies follow neoliberal ideology. Neoliberalism entails policies that create "open," "free," and unregulated markets. Its proponents believe that self-regulated markets result in optimal social outcomes, so long as businesses are kept in check by competitors and individuals are empowered to secure their own welfare. This of course requires government to enforce private property rights, ensure open markets and trade, and show restraint from providing welfare. Neoliberalism means that the state's only job is to both orchestrate markets and to defend market outcomes as accurate and legitimate[4], but in practice, governments usually do much more.

If you have never heard of neoliberalism, you are not alone. It often manifests as an underlying outlook shaping the entire range of policy options rather than a formal policy playbook for one party or interest group. Neoliberalism has been embraced globally and is thus realizing its vision of open flows of money and investment across borders. But as ubiquitous as it is, neoliberalism is a club with an elaborate secret handshake but no membership directory. Economist and philosopher Philip Mikowski writes, "Self-identified neoliberals are hard to come by. There is no political party or national regime that touts the "neoliberal" moniker. It does not denote a professional position in economics or anywhere else."[5] Neoliberal regimes employ a vast array of cookie-cutter economic policies that use private ownership, new markets, and new financial products to address social problems. American electricity markets are the turf on which several neoliberal polices have changed the playbook for businesses that generate and transmit electrons. To fully understand why wind energy makes some people so angry, it is useful to understand how neoliberal renewables incentives work.

As a global macroeconomic phenomenon, neoliberal policies have grown wealth, but there can be no illusion that wealth, or liberty, is rippling out across the market equitably.[6] As geographer David Harvey observes, evidence shows that neoliberal policies usually defend pre-existing inequities. Obviously, those with political and monetary power to start with are best positioned to benefit from playing a game with fewer rules and no referee.

Thus, most neoliberal policies aid the biggest, richest institutions, such as companies allowed to monitor and report themselves, or large investors who realize economies of scale when taxes and fees are relaxed. As executed, neoliberalism is antithetical to the belief that the state should maintain a level playing field to ensure that capitalism is actually inclusive, markets are actually competitive, and improvements in welfare are truly widespread.

Increasingly, citizens are abetting neoliberal priorities when protections once provided by the state are outsourced to consumers under the premise of personal choice. Examples include purchasing safe and nutritious food and reducing one's ecological footprint. In the United States, these are matters of personal preference rather than guarantees for all Americans. A famous case now circulating in the media is the tale of how the plastics industry funded the "Keep America Beautiful" campaign in the 1970s, which essentially shamed consumers for littering.[7] Putting the onus for plastic pollution on consumers served American producers by absolving them of responsibility for their products once they leave the shelf, in sharp contrast to packaging laws in Europe. Community and environmental health, two things we all pay for collectively in the long run, have both suffered. Apparently, humans may ingest as much as a credit card worth of microplastics every week due to the volume of plastics in the environment.[8] A report funded by the plastic industry itself estimates the cost of plastic pollution borne by society at $209 billion by 2025.[9] The point is, emphasizing consumer responsibility is a neoliberal strategy to unburden business. We know from climate predictions that relying on consumers to buy solar panels and new lightbulbs instead of overhauling energy systems also places clear industry benefits ahead of tremendous societal costs.

In what follows, I examine three shifts in the electric industry that neoliberal policy has triggered: (1) market deregulation, (2) commoditizing wind turbines as financial instruments, and (3) creating markets for RECs. I then consider the political and economic impacts that these shifts have had on electricity in Northern New England, arguing that unrelenting focus on profits only elevates the issue of price in the renewables conversation and stifles the issue of climate. That is to say, conservatives and liberals alike fall back on arguments for cost-effectiveness when fighting for or against their preferred method of electricity generation, which leads to the conclusion that expensive renewables are not worth pursuing and reinforces a deeply entrenched zeal for cheap energy.

Deregulation from the Consumer Perspective

The most influential neoliberal electricity policy in the United States has been a steady freeing of electricity markets from what were once very strict protections for monopolies, resulting in a more welcoming environment for renewable energy. In the early twentieth century, electricity generation and distribution were so expensive, so profitable, and so corrupt that on the heels of the Great Depression states granted electric utilities monopoly status to promote stability and incentivize companies to reach more homes and businesses. Within fifty years, distribution infrastructure had been installed coast to coast, and electricity was cheaper, but society faced a desperate need for innovative domestic sources. Amid the nation's second energy crisis in a decade in 1978, federal legislation known as PURPA (the Public Utility Regulatory Policy Act) began deregulating national electricity markets by ending vertical integration, forcing the monopolies to purchase power from independent generation facilities. This legislation opened the door for more renewable power generation at multiple scales. Competition was supposed to drive prices down, and give new electricity sources a fair chance.

A neoliberal policy wave in the 1990s compelled fifteen U.S. states to push deregulation even further. These states, which included Maine and New Hampshire, opened their electricity markets to competition at the level of the consumer. Under deregulation, electricity distribution is unbundled from electricity generation, allowing power lines attached to customers' households to be neutral pathways through which the consumer's preferred energy flows. This unbundling allows consumers to select from a number of potential electricity sources, empowering them to demand renewables. Thus, deregulation in progressive places makes it enticing for renewable energy generators to sell green electricity. In practice, this does not necessarily expand the use of renewables; rather, it creates a way of profitably distributing renewable electricity to customers who are willing to pay a premium for it.

In states where regulated electricity still exists, like Vermont, everyone buys the electricity mix made available to them by the monopoly. State regulators, therefore, are the sole arbiters of environmental progress and customer satisfaction. Monopoly utilities need state approval for their prices, as in most states, but under regulation utility profits are calculated by adding a preapproved percentage onto their operating costs. With no competitors to drive down prices, state regulators are the only downward force on the costs that customers pay. In the simplest of terms, regulators ensure that monopoly utilities do not spend indiscriminately just to squeeze more money from their customers. With ecologically progressive voters, Vermont regulators have the legislative and popular support to also enforce environmental standards, which they do vigorously.

In approximately 30 U.S. states, in regulated and unregulated markets alike, renewable energy is prioritized under legislation that mandates the minimum percentage of carbon-free or carbon-neutral electrons sold by electric utilities, with escalating mandates into the future. These policies are known as renewable portfolio standards (RPS). In regulated markets, consumers can purchase green electricity only if their local utility sells them green electricity. In a regulated market with an RPS, like Vermont, utilities are required to sell green electricity, which means that the costs of renewable electricity are distributed across all ratepayers. Vermont is proof that regulated markets can meet renewable goals because ratepayers shoulder the costs together.[10] In such an environmentally progressive place, this arrangement probably suits most ratepayers. Although I will scrutinize Vermont's unique and highly criticized approach later, it is seen by many as a model of environmental progress.[11]

Deregulation and the federal laws that allowed it to commence have been a boon for newer, smaller renewable energy interests. Hess writes, "within the spaces opened up by restructuring, some interesting opportunities have been created, perhaps inadvertently, for new forms of redistributive politics to emerge."[12] Hess is referring to policy changes that opened energy markets to smaller and smaller energy producers, which in some cases are guaranteed price floors for the electrons they supply to the grid. More innovative still are models of community ownership, solar shares, and solar bonds that now allow electricity rates to be reinvested more directly into renewable infrastructure by removing the need to generate profit.

Hess and other scholars have taken an interest in redistributive energy policies because they serve multiple social and environmental goals. Widely distributed benefits from electricity generation decrease opposition to it. As we know from copious wind research, owning a stake in wind turbines makes neighbors far less likely to find fault with the turbines. Shared ownership also makes it easier to stomach paying more, if consumers know that what they are chipping in is not growing revenues for distant corporations. In the best scenario, shared ownership can both promote efficiency and build community—two strong arguments for wresting control from monopoly utilities.

My informants and the broader progressive public in Northern New England do not explicitly question deregulation or a lack thereof, but it is clear that they care about aspects of both regulated and deregulated markets. Vermont's regulated monopoly, Green Mountain Power (GMP), emulates its customers' eco-consciousness and is assisting in the state's drive to reduce reliance on fossil fuels. It was deregulation, however, that gave rise to small-scale and community ownership models, which are the logical opposite of the much-maligned large projects owned by multinational corporations. However, Vermonters do not seem to clamor for deregulation. And discussion of a state buyout of Maine's electric utilities, a sort of re-regulation, raises alarm bells for some there. In both states, electricity prices are relatively high and expected to climb, and as a result, critics routinely take aim at both utilities and regulators alike. It would seem as though the public in both Maine and Vermont pays heed to the "the grass is always greener" adage, appreciating that a system overhaul may not be in consumers' best interest.

Deregulation from the Producer Perspective

Home to both regulated and deregulated states, Northern New England is a microcosm of the varied forms of neoliberal energy policy that exist nationally. With major wind projects in all three states, it is clear that some aspects of the energy business have little to do with how the market is structured. This is neoliberalism's indelible imprint. David Hess observes that even climate awareness has not displaced profit's central role in shaping state policy. Despite new corporate models and renewable innovations, Hess observes that "the electricity industry as a whole remains structured to enable the continued accumulation of profits by the large IOUs and generation companies. . . . When one steps back and looks at the industry as a whole, there is a mixed regime of hegemonic social liberalism and hegemonic neoliberalism that provides opportunities for capital accumulation by the energy elites."[13] "Hegemony" equates to unquestionable dominance, so Hess is arguing that electric companies can profit in all regulatory regimes because those regimes are engineered to harmonize with companies' business strategies. Hegemony is particularly evident in the way neoliberalism has shaped New England Electricity prices behind-the-scenes.

Even though Vermont's electricity market remains regulated, neoliberal ideology affects the state's utilities thanks to neighboring states that opted for deregulation. This is because the six New England states share a common electricity grid managed through a "Regional Transmission Organization" known as ISO-New England. RTOs took shape in the deregulation era as the region's monopolies broke into entities of varying size and scope. Seven RTOs operate in the US, generally in states that deregulated their energy markets, or in some cases, where grid management was to be significantly impacted by other states' deregulation.

ISO-NE is staffed by electrical engineers, computer programmers, economists, and policy experts tasked with maintaining a reliable and optimally functioning electricity grid, which requires a perfect, up-to-the second balance of electricity supply and demand. ISO-NE must also ensure cost containment and equity among electricity generators so that customers pay affordable rates and a diverse portfolio of generators can profit enough to stay in business and keep the lights on. These tasks are tall orders on their own, but in a region with several new electricity sources, merging via six unique policy tacks on a grid built generations ago for monopoly management, they are herculean tasks.

Vermont's regulated monopoly utilities immerse themselves in murky neoliberal waters when they sell their electrons to ISO-NE. As customers see it, the price on our monthly bill covers the cost of the electrons our utilities make or buy for us, plus whatever profit or overhead they claim. But in reality, how utilities arrive at their prices is far more complex. I have neither the space nor the expertise to explain the inner workings of electricity pricing in Northern New England, but I will offer a few brush strokes to help illustrate neoliberalism's indelible mark, even in a state that avoided deregulation.

ISO-NE buys electricity from power plants and relies on auctions to set prices at the point where supply meets demand, but as is typical with neoliberalism, the transparency we associate with auctions gets clouded by an array of policy add-ons. For example, a bidding power plant tells ISO-NE the minimum price it is willing to accept for its electrons in three separate auctions: one for real-time prices at five-minute intervals, one for anticipated generation the following day, and one for overall capacity in monthly increments three years into the future. This offers three ways for power plants to make money. Moreover, none of these auctions simply find where power plant supply meets grid demand because ISO-NE does not pay the rate that power plants bid. Instead, all bidders receive the "clearing price" paid for the highest of all accepted bids. Furthermore, despite its charge to be fuel neutral, ISO-NE follows a rulebook with special provisions that seem to advantage natural gas and renewables.[14]

This spate of complicated auctions and rules shatters the neoliberal illusion that deregulated markets achieve optimal outcomes without extreme intervention. And yet, as energy Scholar Meredith Angwin describes, RTOs are so married to the neoliberal illusion that they usually address problems with existing auctions by layering on more auctions. The problem, Angwin observes through a review of research comparing RTO states with non-RTO states, is that the convoluted neoliberal mechanisms RTOs use for pricing electricity have not clearly advantaged ratepayers like they are supposed to. They have, however, clearly disadvantaged the safety, reliability, and longevity of regional electricity grids while making only dubious contributions to greenhouse gas reductions.[15]

As Hess argues, utilities' objective is always to maximize electricity sales within the legislative framework provided. The particular rulebook used by RTOs is a central part of this framework for profitability, even in regulated states still served by monopolies. Regional grids introduce enormous complexity to electricity markets due to the guiding ideology of neoliberalism and an unspoken mandate to privilege big energy stakeholders. I will explain in this chapter's next section that projecting neoliberal ideology onto complex policy apparatus is only the latest strategy for serving status quo electricity interests.

Financializing Wind: Green Cash Machines

> What you are not fully grasping is that these deals have zip to do with electricity; they are all about the money.

A second key dimension of neoliberal electricity markets is promotion of the free flow of capital across multiple layers of business or, in other words, making wind turbines attractive investments. Investor-owned utilities provide shareholders with dividends, which means that eventually those companies expand in other regions or other countries with the goal of growing profits. But as the case of wind in Northern New England demonstrates, renewable energy policy built around subsidies has created a new conceptual frontier for corporate accounting and strategizing. Simply put, wind became a good investment overnight. This is why people in Northern New England, like the online commenter quoted above, feel scammed by wind energy.

Tax incentives comprise the bulk of federal wind subsidies.[16] The most enticing incentive is the federal production tax credit (PTC) granted for owners of renewable energy projects. The PTC allows for a 2.1-cent annual deduction per kilowatt-hour generated throughout the first ten years a wind project is online, and it can be banked for up to twenty years in the future. Under President Obama's stimulus plan, the deal for wind developers was even sweeter. Developers could receive a 30 percent tax credit in lieu of the ten-year PTC or, alternatively, a 30 percent Treasury Department grant.

The other major incentive for large-scale wind development is the Modified Accelerated Cost-Recovery System, which permits renewable energy infrastructure to be fully depreciated over a five-year period instead of the typically longer periods. In essence, this means that an entity that owns wind turbines can realize what is for other assets several decades of tax reduction over only five years. Given the enormous expense of ridgeline wind projects, the full benefit of tax deductions and accelerated deprecation is useful only to owners with very high tax liability. In fact, the tax incentives cannot be used to decrease a company's tax bill below 75 percent of its regular tax liability. Because many wind developers are not large enough to carry a tax burden that can fully utilize the incentive, they often divest of their projects shortly after construction, as the tax benefit can be transferred to new owners within a ten-year window.

These policy arrangements explain why large investment banks have both participated in wind development and purchased wind installations after construction. Major investment banks have built portfolios of renewable energy projects across the United States, as the tax burden associated with the banks' enormous profits can be significantly reduced by purchasing "tax-advantaged" investments like wind farms. Offering particularly strong evidence of electricity's neoliberal turn, a new type of fully independent corporate subsidiary is buying wind projects to offer investors low-risk returns thanks to the long-term energy purchase agreements that utilities sign with developers. Known as "Yieldcos," these entities typically distribute a large portion of revenues to investors in the form of dividends. They are the functional equivalent of REITs, or real estate investment trusts, which allow investors to own shares of real estate that pay consistent returns from rental income or timber sales.

Neoliberalism drove industry consolidation in Northern New England states because it allowed larger entities to profit from wind turbine development. With the production tax credit, rapid depreciation that conveys to subsequent owners, and bundled ownership options, the number of entities eager to own wind turbines grew quickly. Turbines became novel opportunities for investors, compelling new business partnerships, acquisitions, and alignments. These opportunities help to explain why Congress and President Donald J. Trump, a vocal detractor of wind turbines, have maintained federal incentives. The world's largest investment banks are seeing enormous tax liabilities disappear by parking their capital in wind projects guaranteeing steady returns.

Making Markets

At Christine's house overlooking the chickens, the stream, and the solar panel, she pulls up on her computer the ISO-NE website showing the current fuel mix on New England's electricity grid. Back in 2014, renewables on a July day accounted for about 5 percent of 21,000 megawatts. I asked if that included the entire region or just the state, and Christine chuckles. She says, "Our load is about 830 megawatts. We are a blip." Pointing out the window to her single solar panel, she remarks that the whole state could probably meet that demand with technologies like the one in her yard. She continues: "But now there's huge uproar about solar, and in particular Rutland, because GMP is going to make it the 'solar capital.' Well, what does that mean? It means that all these out-of-state companies come in. . . . They get the standard offer, what is it, 20 some cents per kilowatt-hour. They make huge profits from their investments. GMP sells the renewable energy credits out of state. They're proposed in neighborhoods, where people are like, well, what's in it for us? Are we getting any electricity?"

Christine is of course describing RECs. RECs epitomize neoliberal electricity policy because they are an invented financial product that necessitated an invented marketplace. State policies make RECs both necessary and tradable, which has provided electric utilities like GMP with a new mechanism to profit in the electricity business, but leaves communities feeling exploited and some environmentalists scratching their heads.

Each kilowatt-hour of carbon-free electricity that is generated has a REC attached to it. The Regional Greenhouse Gas Initiative (RGGI) is a "cap and trade" compact under which eleven states in the Northeast agree to cap their emissions from electricity generation and through which they are permitted to trade their excess reductions to other states. Instituting renewable portfolio standards in RGGI states created a market for RECs, because it helps utilities comply with renewable mandates by trading RECs across state lines. "Clean" generators with extra RECs can sell them to "dirty" producers.

Wind turbines are a more profitable technology for the electricity sector thanks to the REC market, but many people feel as though selling RECs out of state continues a long tradition of selling out to exploitative interests outside the region. The following are representative examples of how online commenters talk about renewable energy projects enabled by REC trading:

Iberdrola will sell the RECs to out of state entities for about 6 c/kWh, so THOSE entities can claim the feel-good halo of saving the world. . . .

Vermont cannot claim anything towards its 90 percent renewables by 2050 goal. All it gets is: Permanently destroyed ridgelines, Disturbed people, and social unrest in nearby towns for a period of at least 25 years, or more.

The purpose of industrial wind in Vermont is to produce RECs to sell to Connecticut so that the Foxwoods parking lot can be kept lit all night with green electricity.

The REC market has functioned as a strong catalyst for wind projects by allowing developers to leverage southern New England's renewable portfolio needs as a stable revenue stream. Thus, the incentives behind ridgeline wind include not only what developers can expect in the rural states where they build but also their anticipated REC profits from utilities across the Northeast. Recall that RECs are the reason Ralph used the term "West Virginia of wind" to describe his take on Vermont's pro-wind policies. Interstate REC trading agitates distaste for outside influences in the rural region, but understanding why wind is unpopular in Northern New England requires diving even deeper into REC trading.

The Internet of Green Electrons. When an electron enters the electricity grid it becomes indistinct and untraceable. The entity that generated it can record its release, but tracking it to a destination is not feasible. So, now that we live in an era of "good" green electrons and "bad" carbon-backed electrons, we rely on generators' records to know the general proportion of good and bad electrons on the electricity grid. Tracking green electrons is accomplished by following the exchange of RECs between power generators and institutions.

In this electricity reality, the benefits associated with renewable energy can literally be traded away. This means that wind and solar energy projects should be thought of as dirty, carbon-releasing infrastructures if their certificates are sold to other entities. Put differently, owning RECs is what allows a company to claim lawfully that it has consumed renewable energy, regardless of how it happens to generate electrons. A coal plant holding RECs can proudly claim to be green.

In October 2017, I attended the annual Renewable Energy Markets (REM) conference to learn more about RECs and how the electricity industry has embraced this new and very neoliberal currency of green electrons. Reflecting the neoliberal mold, a financial architecture continues to emerge from the renewable energy industry to facilitate REC trading. The conference was small, collegial, and progressive in its mission and convened a variety of players. Listening carefully, one could hear the attendees embracing diverse orientations toward the future of electricity. Some looked down the pike at greater electricity democratization, while others clung tightly to their monopoly status. Apart from wanting a renewables transition, what unites them at a singular conference is a shared interest in the financial apparatus through which the monetary rewards of renewable energy are channeled and accumulated. From what I observed at the conference, REC trading is the domain of aspiring financial middlemen of the renewables sector. They are not necessarily energy technocrats or financiers, but rather a new stratum of enviro-traders.

As states and provinces across North America have ratcheted up their renewable requirements, the REC market has expanded rapidly. RECs are purchased voluntarily by residential customers in deregulated states and under mandate by utilities in states with renewable portfolio standards. The distinction between renewable electrons and RECs makes it legally imperative to ensure the validity of the "renewable" label. The entire premise of one workshop that I attended at REM 2017, and a central goal for much of the professional renewables community, is articulating the legality of claims made about electrons' "environmental attributes."

Several conference speakers emphasized the rhetorical strategy of stating that a REC *is* renewable energy and that to adequately grasp REC trading, one needs to approach the concept "from a legal perspective, rather than an engineering perspective."

RECs are traded nationwide, in local, regional, and international REC markets, and not just to pass along renewable attributes or green bragging rights, but because differences across jurisdictions ratchet up profitability. The price for RECs varies wildly across places because some places specify that a minimum proportion of RECs purchased by utilities are local. For example, in Washington, D.C., the 2018 price of one megawatt-hour (MWh) REC was around $400, nearly twice as much as in Massachusetts, the second most expensive market. This enormous disparity in REC values facilitates selling RECs where they are most valuable and purchasing replacement RECs where they are cheaper, a practice known as arbitrage. To arbitrage RECs is to maximize a renewable infrastructure's revenue-generating potential by exploiting what amounts to differences in localities' commitment to CO_2 reduction. These differences allow solar panel owners to make money by selling high-value RECs but then purchase low-value RECs to maintain their green bragging rights or perhaps relieve themselves of environmental guilt.

An enormous amount of effort goes into auditing, accounting for, and verifying claims about the origins of electrons fed into the electricity grid, and lately millions of hours are being invested in developing computer infrastructure to make these functions more expedient and automated. There are already more than ten tracking organizations in North America that account for RECs and the most recent figures, from 2016, reveal that more than fifty-one million megawatt-hours of RECs were exchanged by 108,000 parties that year.[17] The organizations that facilitate these transactions is an international network dominated by nonprofits and nongovernmental organizations (NGOs) deploying eco-oriented venture capital. Their work is engineering the economic infrastructure that is now essential to removing fossil fuels from electricity production. This network and its infrastructures represents significant resources expended in the REC trading community to guarantee that those making public claims to renewable consumption do in fact possess the RECs to support their claims. Such expenditures are emblematic of the enormous investments neoliberal policies demand in terms of market regulation, despite their ideological opposition to regulation.

Belying the uniformity of electrons, enormous complexity can be found in the REC trade. Apart from their value in dollars, RECs are levers of environmental change, another kind of value. Some progressive consumers and businesses prefer this environmental value and "retire" their RECs instead of trading them. For example, many manufacturers value the publicity from sourcing renewable energy more than the profit they might receive from selling RECs or allowing their utility to retire them. The same holds true for residential solar owners. If they sell their extra electrons to their utility, they can choose not to hand over the RECs and instead retire them, so that the utility does not sell them or trade them for lower-tier RECs through arbitrage.

Not only do REC values vary across different policy environments and stakeholders, but geography also differentiates one REC from another because of regional differences in the fuel behind electricity generation. For example, green investors value RECs more if they represent greater carbon reduction potential, such as electrons that displace coal as opposed to natural gas. Coal dominates the American Midwest and South, and natural gas the Northeast and mid-Atlantic. These factors, while they may not directly influence REC prices on national and local markets, do explain individual and institutional priorities in making renewable investments. For example, at REM 2017 I learned about a group of Boston institutions committed to purchasing green energy. After careful research, the group invested in a North Carolina solar farm displacing coal electrons rather than New England wind projects with less impressive claims to carbon reduction.

In the simplest of terms, RECs exist because they are a way for society to promote green energy while also creating a new way for entities to leverage their investments in technology to make money. This is the essence of neoliberal market-making. Governments helped established REC trading markets, and those markets privilege electrons generated from clean sources thanks to policies like renewable portfolio standards that make them valuable. However, fundamentally, RECs are valuable only relative to fossil fuel–generated electrons. If fossil fuel electricity were to disappear because we discovered an ideal renewable power source, the REC market would disintegrate. Thus, we should acknowledge that one of the United States' central environmental policies was engineered to be a peacefully coexisting if not symbiotic side hustle to the fossil fuel industry.

No Double-Counting! The wind boom in Northern New England showed that RECs incentivized wind development under both regulated and deregulated electricity regimes. In Maine and New Hampshire, wind developers rely on the REC market for a large component of their revenue but can also secure demand locally because of the states' renewable energy portfolios and the appeal that renewable energy holds for eco-conscious ratepayers. Wind developments in northern Maine are located so far from large communities that there is no pretense that the power will be distributed in Maine at all. Several developments hold power purchase agreements with utilities in southern New England to meet those states' renewable portfolio standards.

Vermont's approach to RECs has been assailed by states across the Northeast. Under Vermont's Sustainably Priced Energy Enterprise Development (SPEED) program, in place from 2009 to 2015, RECs could be sold to other states after they were "counted" toward Vermont's energy goals. In effect, the policy tied the incentives for developing wind turbines to other states' goals, instead of its own. Among wind opponents, it constituted a "double-counting" of renewable energy and inflated ridgeline wind's financial viability. By forgoing a renewable portfolio standard and instead permitting developers to game the REC market, Vermont's SPEED program took a manifestly neoliberal path. Eventually, this policy was seen by Connecticut as a strategic way for Vermont to privilege its monopoly utility, GMP, while shirking Vermont's contribution to regional reductions in carbon emissions. Vermont Law School's Environmental and Natural Resources Law Clinic summarized the problem in a petition to the Federal Trade Commission seeking to halt GMP from claiming that it sold its ratepayers renewable energy:

> GMP is representing to its customers and to the public, through its promotional materials, public statements, and other communications that it is providing its customers with electricity from renewable sources such as commercial wind and solar projects, thereby reducing the customer's carbon footprint and protecting the environment. In fact, however, GMP is selling substantially all of the Renewable Energy Credits (RECs) generated by these sources to out of state utilities in satisfaction of those utilities' legal obligations to meet mandatory Renewable Portfolio Standards (RPS) in nearby states such as Massachusetts and Connecticut. The net result is that Vermont customers are being misled into thinking that they are buying "renewable energy," when in fact what they are getting is "null" electricity consisting of a mix of fossil fuel, nuclear, gas and other "brown" sources of electricity from the regional grid.[18]

By 2015, when the Vermont legislature enacted a more conventional RPS to replace the SPEED program, the state of Connecticut had filed federal charges to compel Vermont to cease with its "double-counting" of RECs. The issue resolved itself when Vermont conceded that SPEED did not entail an energy mandate, just voluntary recommendations, and therefore could not be in conflict with other states' mandates.

Vermont's 2015 replacement to the SPEED policy has also attracted criticism for its expressly neoliberal qualities. Act 56 set tiered utility mandates for different scales and approaches to renewables. For some of those tiers, Act 56 allows electric utilities to satisfy 75 percent of the mandate through REC arbitrage, selling their premium RECs to other New England states and either buying cheaper RECs from Hydro Québec, which other New England states do not recognize and are therefore less expensive, or paying a modest $0.01/kWh "alternative compliance payment" (ACP). The ACP gives utilities flexibility in the event that RECs are too expensive to buy, too cumbersome to generate, or too lucrative to retire in-state. REC arbitrage means that Vermont ratepayers may be using none of the power created on their ridgelines. In fact, the new law incentivizes grid-tied solar customers to surrender RECs to their utilities. Although this surrender makes sense as a first step toward distributed generation, it also seems like a profitable monopoly is capitalizing on its do-gooder customers. It is through REC arbitrage that Vermont ratepayers effectively *increased* their average greenhouse gas emissions from electricity; one report finds that, on paper, this has amounted to a doubling of the state's emissions over the decade spanning 2008 to 2018.[19] The increase is a consequence of REC accounting practices. Green power that has been de-greened takes on the average carbon emissions of the regional grid, which in New England is more than 75 percent nonrenewable.

Conclusion

The permit application for the one-turbine Dairy Air wind project proposed near my house was withdrawn in January 2020. While the developer cited a changing political climate and increased public scrutiny as the reason, it is likely that somehow the project's economic prospects shifted too. It is possible that local resistance necessitated unforeseen legal expenses, but as I have tried to show, there are almost too many financial variables to keep track of in the regional wind economy.

Even a single locally owned industrial wind turbine is complicated because it necessitates a redesign of the electricity infrastructure that moves electrons and, at the same time, engages new networks of economic actors. A small project may be less enticing for potential buyers down the road seeking tax benefits, or it may be more sensitive to volatility in the REC market. The project in question was small enough to quality for a long-term power purchase agreement through the state's "standard offer" program, but there is ambiguity as to whether the utility buying the electrons can sell the RECs. It is also probable that there is insufficient capacity on the grid to accept Dairy Air's electrons, as the area is a well-known bottleneck, thanks to among other things, the big wind projects that came first.[20] Essential to any project is profitability, and for a project with only one turbine, operating margins will be tight no matter what.

Still, it is hard not to ponder what wind resistance in the region would look like if the first projects proposed were all single-turbine joint ventures with dairy farmers. Instead, many wind projects are internationally owned ridgeline projects with a dozen or more turbines. Their timing, scale, and placement are dictated by profits from their status as financial instruments. This new dynamic of electrons flowing between urban and rural communities is part of a broader neoliberal geography growing from the freer movement of investment. It is an emerging political economy of electricity best characterized as a dense matrix of relationships with new nodes of authority, crisscrossing nearly the entire Western Hemisphere. The communities hosting the machinery, meanwhile, cannot always claim the moral satisfaction of carbon reduction.

This is our wind capitalism. Seeing Northern New England's emerging renewables sector as a changing political economy helps answer puzzles that many citizens in the region are now pondering. Why focus on marginal power generators in places where they work only marginally well? Why not prioritize the region's largest sources of carbon emissions: home heating and fuel? Should our renewable energy solutions be more aggressive and more efficient? The short answer is yes, but that was not the path chosen, because it does not sync with the norms and conventions of American policy in general.[21]

Ridgeline wind turbines are key infrastructure in an electricity industry that dove headlong into neoliberal economic logic. State and federal governments selected a path to renewability mapped out by large energy companies through policy mechanisms that made renewable electricity an enticing new frontier for green investing, creating conditions under which turbines could be profitable—even on remote, steep, mountains. Financiers' money, foreign shareholders' money, and industry money found in Northern New England's ridgelines a safe investment promising steady returns.

Northern New England wind helps expose the contradictions of neoliberalism as a pathway to reaching an environmental goal. The government purports to be "hands off," but expresses clear hands-on priorities in funneling subsidies to large companies and banks instead of institutions that place climate goals first. Fossil fuel subsides, still dwarfing renewable subsidies through the year 2020, also make great evidence of this contradiction. Deleveling the playing field is so common under neoliberal policy regimes that it exposes the lofty fictions on which the ideology is premised. Notions of individualism and open competition seem especially preposterous when energy policy requires the government to contort itself to advantage Wall Street.

In his book *Talking to My Daughter about the Economy*, economist Yanis Varoufakis calls this the irony of market solutions like REC trading, writing that "the only reason to adopt a market solution such as this is because government can't be trusted, and yet this solution depends entirely on the government for it to work. . . . The reason the rich and powerful, along with their intellectual and ideological supporters, recommend the complete privatization of our environment is not that they are opposed to government; they're just opposed to government interventions that undermine their property rights and threaten to democratize processes that they now control."[22] Like Varoufakis, who sees democracy as the antidote to economic policies that are putting our planet in peril, I propose that greater citizen participation in energy policy is essential.

The second part of this chapter explores the political side of the political economy equation, elaborating on why citizen involvement in energy is difficult to achieve. I examine the social structures in which neoliberal policies unfold, arguing that energy industry players and government officials are locked in an antiquated model of electricity planning and regulation that works against climate change mitigation by maintaining emphasis on costs. The example of residential solar illustrates how climate change mitigation butts heads with industry survival strategies, making an aggressive, efficient transition difficult to implement, even in progressive places.

Part II: The People

A Solar Diversion

On a chilly, wet November night in 2014, I attended a public hearing about solar energy in Maine. Maine's Public Utilities Commission (PUC) was seeking input on a proposal by Central Maine Power (CMP) to reprice its net-metering contracts with solar owners. CMP is Maine's largest electricity distributor, an investor-owned utility (IOU) that is part of Spanish company Iberdrola's North American holdings. Like electric utilities across the country, the precipitous drop in the price of solar panels was showing up on CMP's balance sheet. CMP argued that the price it was mandated to pay solar owners, known in the industry as the feed-in tariff for net-metering, was becoming unsustainable as more and more households installed solar panels. Reflecting on the dreary weather, I could not help but wonder whether CMP underestimated Maine's solar potential when net-metering was first introduced. The company does not underestimate that potential anymore.

The room was overflowing with Mainers testifying against CMP's proposal. While the attendees' average age was lowered by college students donning yellow T-shirts in solar solidarity, most of the homeowners offering testimony looked well over fifty. Many cited a fixed income as the reason they had invested in solar, because it would ensure fixed energy costs in retirement. As usual, the parking lot seemed dominated by late model Volvos, Subarus, and hybrids.

I stood half in the doorway as several dozen homeowners approached the microphone to speak. They testified that the rates CMP had promised them were what made their renewable energy investments feasible. Several offered a financial accounting of their monthly expenses and posited that reducing net-metering rates would harm their budgets. For the many senior citizens offering testimony, stable electricity rates afforded them a sense of security. And several who testified reminded listeners that carbon emissions should be the primary currency of concern for energy policies. These speakers argued that curtailing incentives for solar electricity wastes the carbon-reducing potential of the expanding solar industry.

It would seem that CMP, like electric utilities across the United States, took note of the high-end cars in the parking lot that night. The narrative that electric companies are using to renegotiate net-metering rates depicts solar owners as an affluent minority who reap benefits at the expense of poor customers who are left to shoulder a growing proportion of fees for utility lines, poles, and substations. This narrative gains easy political traction in a society with wide and expanding inequality, in which volatile swings in energy costs place a higher burden on poor households. CMP's argument ultimately resonated with Maine's Public Utilities Commission in 2015. Net metering was replaced with a law in 2017 that added substantial new burden to grid-tied solar, causing new installations to plummet. Net metering did not return to Maine until 2019.

Recent studies of net-metered solar electricity raise doubts about the claims for equity made by utilities. Utilities argue that net-metering obligates them to overpay for grid-tied solar systems, but the veracity of that claim can be assessed only by considering *when* the electricity enters the grid. Electricity prices are determined by balancing supply and demand at five-minute intervals, which means that when demand is highest, electricity generators can charge higher rates per kilowatt-hour. Because solar panels tend to produce excess power when it is most expensive—during hot, sunny, daylight hours—the uniform net-metering rates paid by utilities to solar owners may be nowhere near as damaging for utilities' bottom lines as they claim, because residential solar can erase the need for expensive and polluting peak-generation apparatus. A 2018 analysis of a July heat wave in Vermont confirmed this benefit, and a 2020 study places the benefit at over a billion dollars across the region.[23]

Irrespective of the precise value of solar-generated electrons, the sudden concern for ratepayer equity has provided utilities nationwide with the political ammunition to lobby for lower rates for solar owners, slowing the solar transition by reducing the number of households for whom the transition makes financial sense. Although net-metering policies exist in more than thirty states, several states have made rollbacks, including Maine and Vermont. New Hampshire, like many conservative states, has been unable to expand programs because of intense pressure from national energy lobbyists.[24] Fundamentally, the utility-led backlash against net-metered solar panels means that the pace of transitioning away from fossil fuels is slowing to the rate at which electric utilities can be profitable, regardless of the imperative to address climate change.

Utilities' efforts to reduce net-metering rates show the tension between climate and profit embedded in policy and political rhetoric. At the heart of this tension is the grave threat that small-scale renewables pose to existing power companies. Reducing feed-in tariffs is part of a suite of policies typically associated with petro-states like Oklahoma, where influential oil interests maneuver for dominance.[25] Such policies discourage critically important low-carbon energy and coddle corporate dinosaurs of the electricity industry that are already ignoring grid maintenance thanks to the pressures of neoliberalism.

To be sure, the electricity grid is something that should concern us. Anthropologist Gretchen Bakke argues that it is antiquated, decrepit, and growing perilously more vulnerable to ice storms, tree limbs, and squirrels by the day.[26] Energy Scholar Meredith Angwin attributes this decay in part to the recent onslaught of renewable policies, observing that by separating policy makers from any direct control of the electricity grid, deregulation emboldens legislatures to set energy goals without having to worry about the grid's capacity to sustain those goals.[27] From this perspective, policy makers are like doting grandparents filling grandchildren with caffeine, knowing it is not their responsibility to put them to bed. All of this is to suggest that neoliberal policies have introduced real structural impediments to maintaining the grid, but residential solar owners are just convenient scapegoats.

Utilities know that massive changes are underway thanks to advances in solar and battery technology. They know a centralized electricity grid may not be necessary for much longer. This prospect, referred to in energy circles as the "utility death spiral," instills fear into electricity companies. If there are fewer customers on the grid, those customers would pay higher prices per kilowatt-hour, so they would probably consume less electricity or finance solar panels, lowering utility profits even more. In rural regions populated with eco-conscious retirees, mass detachment from the grid is conceivable. One analysis of Michigan's Upper Peninsula, a demographically similar region to Northern New England with even higher electricity rates, predicted that the demise of legacy electricity companies is near. The authors estimated in 2016 that, by 2020, 92 percent of seasonal homes and 75 percent of year-round homes would save money by installing solar and detaching from the local grid. Further, their analyses projected that 65 percent of all owner-occupied households will be able to meet their electricity demand with solar battery systems *and* afford to purchase and install the systems by 2020.[28] Standing in the way of those predictions coming to fruition are efforts by Michigan lawmakers to put the brakes on solar installations.

Back to Wind: The Industry's Safe Bet

The plight of residential solar customers brings Northern New England's changing political economy of energy into sharper contrast because it lays bare the close and collegial relationship between utilities and governments. How in the throes of worsening climate events are electric utilities able to slow the pace at which we convert to solar? The short answer is that the energy industry may not be driving the climate policy bus, but they have one hand on the steering wheel, and they always have.

Comparing ridgeline wind to residential solar reveals key differences between a technology that is synergistic with society's centralized electricity model and a technology that threatens that model. Northern New England's ridgeline wind turbines conform to utilities' old habits and promote lucrative new habits. Environmental sociologist David Hess argues that the energy industry incorporated wind technology into its portfolio when wind "could be utilized for grid-based production." Smaller turbines, Hess argues, "become marginalized as capital investments have flowed into designs that are more compatible with existing technological systems and investments."[29] In other words, grid-scale wind turbines are ideal for large electricity interests, distracting from research and design for smaller-scale versions. They enable utilities to meet renewable mandates without changing much. In contrast, solar technologies benefit electricity users at virtually every scale and therefore threaten utilities' existence. Whereas residential solar is a threat to matrices of corporate and political power, ridgeline wind turbines merely extend the umbrella of global industries that profit when ratepayers turn on the lights.

Like efforts to curtail solar panels, efforts to promote ridgeline wind exemplify how public-private collaboration is shaping the country's renewables transition. This part of the chapter focuses on those collaborations, characterizing them as a network of long-established bureaucratic and regulatory structures. I describe the diverse interests united in the wind industry as participants in an evolving schema of decision-making that, despite its green veneer, is still just an industry lobby. This chapter recounts historical research showing that regulatory structures are designed to referee a game with predestined winners and losers. The referees of this game, the regulators, are the targets to which wind opponents aim their protest. But rather than focus on referees, it is in society's interest to focus instead on the rules of the game. The two central objectives of our electricity game are the antiquated goals of cheap electrons and the maintenance of privileges and protections for the companies that provide them.

The Origins of Influence: Northern New England's Eco-political Culture

Starting around the year 2000, Northern New England states, and Vermont in particular, nurtured alliances between lawmakers and green businesses working together to build renewable energy policies. This momentum fell squarely within a regional political project built around the rhetoric of green jobs. "Green jobs" are jobs in enterprises that minimize their ecological impact, address environmental degradation through innovation, or both. The phrase is now a climate mantra throughout the world and rose to national prominence in 2009 under President Obama's stimulus plan, but the idea has long had traction in the progressive politics of Northern New England. In contrast to states that attract large private employers in the manufacturing sector, the region's relative underdevelopment made even modest numbers of new jobs appealing, and there was no dominant polluting industry in the region to take offense at the inference that their jobs were somehow dirty.[30]

The push for green jobs has coalesced alongside a green-voting constituency as it behooves individuals to act politically in accordance with their local industries. Even in a home heating sector dominated by petroleum, many Northern New England businesses have grown their energy portfolios to include alternative energy sources, including solar hot water and biomass. From the region's green jobs movement came the emerging sectors of green building and design, eco-products, and renewable energy. The products and services of this green economy include naturally sourced and biodegradable consumer products, organic foods and goods, solar installation, weatherization and heating systems services, and renewable energy services. A 2017 report by the Union of Concerned Scientists ranks Vermont highest among U.S. states in the number of green energy jobs per capita, as a result of large numbers of people being employed in energy efficiency and solar installation. Maine ranks in the top ten nationally for its proportion employed in the wind energy sector.[31]

The ascendance of Northern New England's green jobs agenda synchs with a larger capitalist narrative articulating that business growth need not compromise the environment but can in fact be mobilized as a solution to environmental problems. Wind turbines, along with several other iconic environmental-branded products, sync with the narrative that green commerce can put people to work, save the environment, and still turn a profit for local businesses—political gold among a progressive, rural electorate with few other industries.

It was during the green jobs era that wind developers and ancillary wind energy experts put stakes in the region or entered an expansion phase, well before federal incentives for wind energy took effect. By 2002, Vermont was already home to sixteen wind energy companies, including three manufacturers.[32] The wind industry's local roots and importance to the labor force gave the industry political relevance when the technology, price, and federal policy aligned to make wind turbines viable.

Opportunity Structures Aligned for Wind. National energy policy has at times aligned with Northern New England's favorable orientation to conservation and progressive environmentalism. Environmental sociologists Coley and Hess described these periodic alignments as the coalescing of "political opportunity structures."[33] Energy-related political opportunity structures, the authors observe, can advance from within individual states but typically do so in response to the national political climate set by presidential leadership. During conservative presidential administrations, progressive states have made slow but steady progress in ramping up their renewables policies, and conservative states redouble their investments in fossil fuels.[34] Northern New England's wind energy industry has benefited from consistent progressive politics in the region, but the industry's capacity to install large wind turbines in the region relied on national and state opportunity structures aligning in the form of federal subsidies and state statutes.

The Region's wind energy became a convenient focal point for progressive government leaders and a lightning rod for wind resisters. With overlapping interests in wind's success, businesses and lawmakers entered into formal and informal energy coalitions. Those coalitions are essential to understanding the backlash to wind, as the region's wind opponents see the power and authority the coalitions exercise as a form of corruption. Several early editorials reference the "Big Business" and "Big Money" interests poised to benefit the most from wind development.

Mutual Dependence and Close Alliance. By virtue of their close relationships and shared professional interests, representatives of the state form what energy scholar Frank Geels termed "close alliances." Geels identifies three pathways through which private interests shape energy policy in alliances: (1) through personal relationships that garner businesses' access to policy makers, (2) through shared understandings of issue framings that result when policy makers subtly "internalize the ideas" advanced by business, and (3) through direct and strategically targeted lobbying activities.[35] The models of wind energy development deployed in Northern New England must be understood as a product of these pathways, through which ideas pass between private and public interests.

Geels's three forms of alliance usually tip in favor of business interests. Drawing from others' work, Geels elaborates that alliances between state regulators and energy companies are part of a much larger problem with American capitalism. Put simply, close relationships between business and the state exist where it seems they should not because they are threads of equal size woven together to form the fabric of national welfare provision. The electricity industry is not unique in this regard. Geels explains that policy makers form close alliances with business because of their "mutual dependencies."[36] Businesses need the government to make and enforce fundamental rules of the fields in which they operate, such as property rights and regulations governing exchange, finance, and accounting. Policy makers rely on businesses to generate steady economic growth reflected in tax revenues and jobs, which are benchmarks of the policy makers' personal accomplishments and essential goals of the state. Thus, there exists under capitalism a natural incentive for collaboration between public officials and private enterprise, which endows businesses with a dense cache of "structural power" to wield in politics and policy. What is exceptional about electricity is its simultaneous status as a public good, an entitlement, an essential production input, and an environmentally politicized product, which makes the structural power of electricity companies especially potent.

Northern New England's Wind Coalitions

Wind energy coalitions advocated for a piece of the region's electricity pie at an opportune time. Fossil fuel costs were skyrocketing, technical capacity was evolving locally, voters were progressive and ecologically concerned, and federal incentives were one election away. In the deregulated states of Maine and New Hampshire, turbine development showed particular promise because ratepayers demanding clean energy were willing to pay for it, and power could be sold to more affluent out-of-state customers.

Media coverage between 2000 and 2005 shows that proponents used wind power's long-term price stability and contribution to energy independence to sell it to the public. As the public debated early ridgeline wind installations in local media, the following editorial (excerpted) appeared in a "point-counterpoint" editorial section of Vermont's largest newspaper: "Wind energy can provide us with some level of price stability. Because its fuel (the wind) is free, wind energy is not subject to price fluctuations. Once installed, the price of wind energy is predictable and stable for at least the next 20 years."[37]

As wind projects began appearing on the region's ridgelines, the number who benefit from the industry expanded thanks to growth in a host of ancillary occupations. Beyond those trained to maintain and service the massive turbines, an army of consultants has spun off from the industry. Meteorological consultants are required to ascertain whether or not a proposed location is suitable for development. Before construction, ecologists specializing in terrestrial and avian fauna must assess site plans and weigh in on improvements to dampen the impacts of wind projects. After construction, acoustic experts, hydrogeological engineers, and ecologists must monitor turbine installations to ensure compliance with standards for noise, runoff, and animal safety. Community dissent has necessitated public relations consultants and meeting facilitators. And of course, attorneys are involved at every phase. This means a growing wind labor force living across New England needs more wind development to secure their next jobs.

The political machinery that gave rise to green jobs now benefits from their expansion. The renewable energy sector enjoys a particular cohesion and proximity to the political process because of its prominent place in the green jobs narrative that has bolstered development platforms and public morale. For Maine, New Hampshire, and Vermont, in comparison with states where major industries are located, the energy sector fills a void in the region's political scene. Tourism and agriculture are dominated by small employers concerned with simple policies that promote market access. By comparison, the renewable energy sector includes global businesses operating at the national level and pursuing creative policy changes to elevate their products and services. Thus, they have a more pronounced need to influence legislation and more money to make friends in the state capitols. For many of those who criticize wind turbines, the green energy sector's political clout has become a focal point.

Energy Alliance in Practice

The Revolving Door

> That Spanish wind developer gave
> ex-Maine governor Baldacci a vice-
> president seat after he royally raked over
> the unincorporated areas of Maine with
> an expedited wind development law
> that stomped all over the people living
> there and has totally polluted the Maine
> mountains with junk turbines.

Membership in Northern New England's renewable energy coalitions is a matter of public record thanks to news media, and wind opponents cite politicians' moves into the private sector in the same way sports fans talk about trades between teams. Without critiquing wind energy explicitly, news reporting has documented several examples of interlocking relationships between elected officials, the state's renewable energy companies, and nonprofit advocacy organizations. In addition to shaping and advocating for the policies that first promoted wind turbine development, the coalitions have been observed playing a behind-the-scenes role in advocating for relaxed regulation of the wind industry. What follows are the most prominent examples.

Former Vermont governor Peter Shumlin announced his first run for governor in a wind turbine manufacturing facility owned by David Blittersdorf, who, upon Shumlin's victory, was appointed to the board overseeing the state's Clean Energy Development Fund. Newspapers later reported that Mr. Blittersdorf, also a wind developer and campaign contributor to Governor Shumlin, oversaw changes in the fund's distribution policy that resulted in his businesses receiving several million dollars in tax credits from the fund. Jim, my informant who grew up farming in Vermont, recounted the Blittersdorf affair this way: "He was sitting on the board and he voted himself a multimillion-dollar contract and then resigned from the board for conflict of interest." Blittersdorf made headlines in 2017 when his text messages to a legislator were obtained through an open records request. In them, Blittersdorf warned the legislator that proposed turbine noise standards under debate were too restrictive for Blittersdorf's projects and threatened, "I will lose over $1m in development cost to date. . . . I will also need to build the $50m in turbines in MA, RI and NY. Wind in VT will be stopped completely. We cannot get to 90% by 2050 without mix of renewables that must include wind." Blittersdorf's family foundation is a large contributor to the Vermont Public Interest Research Group (VPIRG), the loudest NGO wind advocate in the state, which was criticized in a 2018 editorial: "A sizable amount of money pours into VPIRG's coffers annually from big energy developers and the foundations they support. Several of these energy developers sat on VPIRG's board of directors and helped to lobby state lawmakers to draft the bills, pass them through the Legislature, and streamline the permit process for the businesses from which they now profit."[38]

Vermont's largest monopoly electric utility is an integral player in the state's energy policy and in politics generally. Mary Powell, the now retired CEO of Green Mountain Power (GMP), was a campaign contributor to Governor Shumlin and chaired the committee that raised funds for his inaugural ball. As reported by the Burlington newspaper *Seven Days* under the headline "Tilting at Turbines," several other GMP board members and employees had close ties to the governor and his administration.[39] The same news story reported that the state's commissioner of the Department of Public Service, the individual charged with protecting citizens' interests in energy matters before the Public Utility Commission (formerly the Public Service Board), is married to an attorney in the law firm representing GMP. Robert Dostis, a former Vermont legislator central to the failed SPEED legislation, later accepted a public relations position with GMP. And Chris Recchia, another former director of the Department of Public Service, had previously worked for a biomass energy industry group.

Maine's vast sparsely settled lands make it even more profitable for wind development, which investigative journalists have noted in their coverage of renewable energy coalitions. Former governor John Baldacci also received criticism for maintaining close links to his state's renewable energy industry. Baldacci appointed his friend and former chief counsel, Kurt Adams, to chair the Maine Public Utilities Commission (PUC) in 2005. Adams left the PUC in 2008 and took a position with a national wind developer. After his departure, it was disclosed that Adams had received an equity stake in the wind developer while still serving on the PUC.[40]

Criticism also flared in Maine newspapers around a state legislator who drafted wind legislation while his wife led the law firm representing several large wind companies operating in Maine. Documents obtained through a Freedom of Information Act (FOIA) request in 2014 revealed that wind representatives drafted turbine-siting legislation that would severely limit public engagement in the process. They accomplished this by capitalizing on their close partnership with state senate president Justin Alfond, whose political action committee had received several thousands of dollars in wind industry contributions. The FOIA records revealed that the wind industry coordinated political theater to promote legislation favoring the industry by soliciting a contractors' group to sign an editorial that Alfond published in state newspapers and by organizing supporters to pack the statehouse when the bill was debated. Maine investigative journalist Naomi Schalit wrote, "The story of how the wind industry's problem was taken up by Senate President Alfond, D-Portland, and his staff demonstrates a deep level of coordination between special interests and legislative leaders that often leaves citizens on the sidelines of the democratic process."[41]

In Maine and Vermont, the close relationships between politicians and renewable energy interests have resulted in no formal sanction. To the contrary, the public and the media have been overwhelmingly supportive of wind energy, which led Vermont blogger Robert Maynard to assert that "Greed is OK when it is 'green.'"[42] Maynard's concern is of course that membership in this apparently tight-knit network links green energy policy to political compensation, which could impair officials' capacity to make transparent decisions that withstand deliberative public and scientific scrutiny.

Assailing self-interest in politicians does little to expose the complexity of wind energy's deployment in Northern New England, because elected officials' devotion to campaign contributors has become as intrinsic to American politics as shareholder dividends are to businesses. An alternative perspective suggests that the region's sudden appeal to energy capitalists is simply pushing it in the direction of most other states, where energy interests grease the wheels of politics vis-à-vis coalitions of shared interests. This phenomenon is not unique to Northern New England. A 2015 *Wall Street Journal* editorial on the abrupt resignation of Oregon's Governor John Kitzhaber, titled "Green Love Is Blind," observed: "The modern green machine is a network of wealthy foundations and consultant groups that finance activists who promote and advise sympathetic politicians."[43] In Europe, and in Denmark and Germany in particular, industry weight thrown behind the cause of renewable energy is celebrated as the driving force behind the region's many energy accomplishments.[44]

The national renewable energy industrial complex is important context for understanding wind opposition, but it should be noted that citizens do not read about political back-scratching and then renounce wind energy because it seems corrupt. In many cases citizens experience firsthand being shut out or priced out of the spaces and processes in which energy decisions are made. The following account comes from an online news comment posted by the administrator of a small town who experienced the wind coalition's long reach through the Agency of Natural Resources (ANR) and the Department of Public Service (DPS).

We experienced the horror of a dishonest, scheming developer attempt to put nearly 500 foot industrial ridgeline wind towers right down the middle of our town. What was even more painful was finding out that the developer had been working with the ANR for over a year prior to any notification to the town, the ANR had set up a Wind Committee that involved everyone but the citizens of Vermont. Don't our taxes pay for the ANR budget? The next horror was trying to work with the DPS, "representatives of the citizens," we found very quickly that the citizens . . . were locked out of discussions with the DPS, they represented the developer and the Shumlin administration, then as we entered the PSB hearing at an expense of tens of thousands of dollars, we understood that our government no longer cared about us. Four years later our town is still reeling from the effects of this nightmare and the lawsuits go on. I could go on and on how this struggle was not just with a developer with very deep pockets but included the Vermont government that we had faith in. I can guarantee you that faith no longer exists and these developers and the Shumlin administration's approach to Vermont citizens and communities are directly responsible for the nightmare we see in Trump right now.

Echoing the point made in chapter 2, the wind industry has suffered the consequences of disrupting the region's usually high confidence in government. Wind developers' tactic of aligning closely with public officials resulted in voters rejecting candidates seen as darlings to the industry or political insiders. The commenter makes the astute observation that when voters lose faith in leaders because of one issue, that loss of faith sends ripples across the political landscape, leaving unintended consequences in its wake.

Regulatory Capture

In the domains of rulemaking and the setting of standards for wind turbines, the close-knit relationships between energy companies and regulators receive tremendous scrutiny in Northern New England, and Vermont in particular. While Vermont's pro-wind governor was seen as the loudest cheerleader on the sidelines, it was his appointees on the state's Public Service Board (PSB), now the PUC, who were seen as the players on the field, carrying the ball for "Big Wind" at every play. These are two exemplary comments posted to online news stories:

> The cities and towns of Vermont have had enough of three corrupt people who have worked a Public office into the grip of regulatory capture. The PSB will no longer dictate what industrial rubbish will be blasted into Vermont towns for the purpose of wallet-padding. The biased and utility "bought and sold" PSB only needs to make one decision—to resign now or wait until after the election.
>
> You can teach me nothing about the PSB. I've experienced their arrogance and corruption first hand. When a department of state takes its orders from special interest groups and distorts the law to benefit this group and disregards and disrespects Vermont citizens . . . that's called criminal not democracy.

As these comments indicate, wind opponents target the PUC for what they interpret as a consistent bias in favor of wind developers. Borrowing from an academic lexicon, the first commenter labels this "regulatory capture," which connotes the adulteration of supposedly impartial adjudicators with the goals and priorities of those the regulators are supposed to regulate.

Regulators in all three states face pushback over wind and solar siting and the processes leading up to approval. With few exceptions, regulators decide in favor of energy developers, sometimes in contradiction to their own rules. In Vermont, for example, citizens have petitioned the PUC to enforce its own protocol for approving meteorological test towers and for enforcing runoff management plans. Attendance is sometimes high at the board's public proceedings, and several years' worth of citizen videos circulate on YouTube. The public now sees firsthand that the PUC's "quasi-judicial" process offers its board members generous latitude to regulate utilities without following conventional standards of transparency or jurisprudence.

Many lament that the board's impaneled decision-makers are chosen for their industry knowledge and not their legal training or demonstrated impartiality.

In addition to enforcing renewables regulations, state electricity authorities enter into complex interdependencies with utilities through price regulation. In Northern New England, price has been a consistent thread in the discourse advanced by the business-regulator alliance. Wind turbine developers promise low costs per kilowatt-hour, and the public service boards, following their mandate, privilege affordability over other factors. This presents an incentive for developers to overestimate the generating capacity and underestimate the costs of their proposed projects, which evidence suggests they may be doing in Vermont.[45]

Further, electricity rate changes must be approved by the public utility boards, requiring that public officials act as arbiters of their coalition colleagues' profits. In regulated states like Vermont, profit is a guarantee. Vermont guarantees Green Mountain Power's profits will not sink below roughly 9 percent, a figure that floats with the rate of return on U.S. Treasury bonds. The state keeps this promise to its monopoly utilities by allowing it to pass the cost of its expenditures to ratepayers, so long as regulators approve. This profit guarantee helps explain why a multinational energy conglomerate would want to own a monopoly utility. Owning GMP is like owning a mint that prints money at a guaranteed rate.

It is easy to see why coziness between energy executives and state regulators raises suspicions. Vermonters received specific cause for concern in late 2018 when an anonymous whistleblower at the state Department of Public Service published a troubling account of a colleague's firing. The letter, which was later corroborated by the fired employee, described a culture of cozy relations between state electricity rate setters and Green Mountain Power. Allegedly, the employee was fired for looking too closely at GMP's proposed rates—which, the public is left to infer, the utility calculated using methodology that would not stand up to scrutiny.[46]

News reports have observed that inequities in rate setting takes root in several structural advantages working in the monopoly utility's favor. For example, a public radio story from 2017 notes that GMP enjoys copious time to plan and strategize its rates, while state regulators have a narrow window during which they must vet and approve GMP's rate proposals. If the state cannot assess rates in time, they take effect regardless, potentially costing ratepayers millions. Moreover, the expense of legally contesting GMP's proposed rates presents a cost barrier to the state sizable enough that it has never actually been done. Even Vermont's often maligned PUC acknowledged problems with the state's rate-setting methodologies, writing that public statements made in testimony to the committee "reflect significant disagreements about regulatory policy and the best strategy for protecting the interests of Vermonters."[47]

What is most intriguing about the allegations against GMP is that the utility is recognized nationally as an environmental hero, perhaps the greenest electric utility in the United States. GMP uses solar batteries in customers' homes to avoid firing up power plants during peak demand, and the utility is building expertise to manage the microgrids of the future. In fact, GMP sought and received status as a B Corporation, a distinction reserved for companies committed to social and environmental principles. GMP is also owned in part by Enbridge, a fossil fuel conglomerate with an abysmal environmental track record.

Looking back to how Geels categorizes alliances and pathways of influence, it is evident that the public perceives that the relationships between business and government are direct and informed by businesses' agenda rather than simply a shared interest in feeding electrons on the grid. Public officials might even agree. A state regulator I interviewed made it clear that the regulator's relationship with a major utility entailed a common vision, direct lobbying, and the solicitation and provision of technical expertise. Similarly, after reading the comments posted earlier about the ANR's secret yearlong negotiations with a wind developer, I noticed on the ANR website that the agency seems quite proud of its program to shepherd wind developers through the siting process. From the ANR's perspective, secret energy planning is no secret at all; it is a public service that the agency is happy to offer.

The force field enshrouding energy planning persists at the regional level as well. In her 2020 book "Shorting the Grid," energy scholar Meredith Angwin paints a troubling portrait of the illusory channels through which citizens can impact energy priorities and practices overseen by ISO-NE. Angwin observes that New England Power Pool (NEPOOL), the consortium of electricity insiders that birthed ISO-NE, still plays an active role in directing it. NEPOOL leaders sign off on ISO-NE leadership but have no responsibility to transparently disclose, much less sever, their own interests in energy companies.[48] Further, NEPOOL charges membership fees, affords electricity users a miniscule say (16 percent) in voting, and closes all meetings to the public, including journalists. New England governors, trying to shepherd hard-hitting greenhouse gas reductions, have lashed out at this secretive organization and its influence on ISO-NE. One newspaper article carries the banner: "ISO keeps your lights on. Just don't ask how," and quotes New Hampshire's Consumer Advocate, D. Maurice Kreis as saying, "The ISO is unaccountable and rogue."[49]

The crux of the matter is that today's public-private relationships impact both the environment and the public's trust in democracy. It is clear that modern energy regulators and energy developers exist in a symbiotic professional space, collaborating on energy projects of great importance to our future with little concern for whose ideas are plotting the path. But it would seem that the industrial wind coalition, at least in Vermont, has similarly little concern for transparency, and is ignorant to the political consequences.

Shane, the environmental reporter I interviewed, tells me that talking to wind developers is difficult because of the lengths to which they go to cloak their plans in secrecy: "Developers on wind, they don't want to, how should I say it . . . disturb the project's viability in any sort of way, so the easiest way to do that is to keep them on hold, keep things quiet. Some of the developers I've spoken with have said that a lot of work goes on behind the scenes before these projects become public, and that's to ensure that if we're going to make these huge investments that everything is . . . 'All your ducks are in a row.'" It is clear to Shane that developers calculate with precision the timing of their projects' advancement through regulatory phases to minimize the public's opportunity to intervene. Developers do not want the public to know where they own land or what the status of their development plans are. Shane told me that he has resorted to multiple means of sleuthing in the course of his reporting because he sympathizes with the anxiety that comes from communities' uncertainty about pending wind development plans.

Lessons from the Past

For historians of American electricity, perceptions of corruption are old news. That politicians, regulators, and developers work in coordination to secure electricity through mutually beneficial arrangements is more a historical fact than a suspicion. Electric utilities power virtually all industries, and therefore entire economies, which gave rise to unique advantages in shaping the policies meant to regulate them. Electricity governance today is the legacy of the advantages established long ago.

In his history of electricity, Thomas Parke Hughes[50] points to the scalable profitability of generation technology as the reason utilities first aggressively convinced households to consume electricity with abandon. Residential consumption in the evenings balanced industrial consumption during the day, maximizing utility profits. Companies have worked for a century to ensure that their product will be in constant demand and that states will be under public pressure to ensure demand is met. Not only has the electricity industry been resolute about maximizing the capacity the industry has, it has justified expanding capacity using sketchy economic models based on its past success turning U.S. households into electricity addicts. More volume begets more sales, which result in higher profits. As electricity became essential for society, electric utilities were amassing immense political power.

Utilities' political power has always been concentrated in their command of the criterion by which they are regulated. After a few disastrous decades of swindling and greed by the first electricity millionaires, the government's remedy was the one companies pushed for: monopolies. With the passage of 1935's federal Public Utility Holding Company Act (PUHCA), states awarded monopoly territory to companies and established regulatory boards to keep companies in check. While those boards may seem like arms of the state, historian Sharon Beder describes them as structures designed by utilities themselves to deflect criticism for receiving monopoly status in the first place. These regulatory boards were stacked with appointed energy insiders who catered to utilities looking to expand capacity at customers' expense.[51] These are the same public utility boards and commissions that exist today.

For as long as electric utilities have been regulated in the United States, regulators have been committed to keeping utilities happy by promoting growth and to keeping customers happy with low prices. A disproportionate amount of electricity was usually going to major industries at prices set low enough to keep those industries from just building their own power plants, leaving residential ratepayers to shoulder the burden of expansion. Ordinary customers felt entitled to low prices because regulators promised them low prices, but regulators set residential rates that allowed expansion. These arrangements are the reason electricity is expected to be both cheap and abundant and why regulators who seem to put utilities' push for abundance before customers' push for cheapness are seen as corrupt.

Sharon Beder's book *Power Play* makes it clear that electricity industry tactics were dastardly and insidious, going far beyond corrupting regulators.[52] Beder details how power companies funded shrewd public relations machines to prop up an illusion of regulation even as power companies called the shots through their influence of public boards. Monopoly utilities also spent untold fortunes fighting against municipal takeovers, funding think tanks, distributing curricula for schoolchildren, and paying salaries to university faculty. Some orchestrated covert marketing campaigns, enlisting homemakers to host dinner parties in which they parroted scripted talking points that stoked fears of communism.

The industry fought deregulation vehemently, and although they failed, many power companies that lost their monopoly fought successfully to pass old costs onto future ratepayers. In other words, power companies were allowed to keep the spoils they had plundered from decades of unchecked expansion. The most egregious part is that for decades the costs of their deceptive public relations were tacked on to companies' operating expenses, which means that until corporate watchdogs started noticing, ratepayers were unwittingly funding utility propaganda. While I do not present this history as an indictment of contemporary oversight in Northern New England, it shows that state utility boards and commissions across the United States represent a deeply entrenched ideological status quo oriented toward expanding grid capacity on the backs of residential ratepayers.

Rural places generally followed less contentious, but also less scrutinized, electricity evolutions. Small, remote communities formed cooperatives allowing members to coordinate and oversee the electrification of their own homes and farms. While cooperatives are democratic in structure, their governance and leadership has frequently devolved into rule by small de facto cliques who have the same incentive to ingratiate themselves to state regulators as private power executives and follow the same allegiance to keeping costs minimal.[53]

Productivism and Price

For Ralph, who sees mountains as ecologically and spiritually invaluable, the electricity industry and the consumer society it supports are culpable in building a power system without limits. Ralph told me, "The goal these days is to provide as much electricity as people want, unlimited supply, and we do that by building more and more generating capacity. I believe that

that is a perversion of the human spirit and will ultimately lead to a failure of planetary systems." The snowballing of electricity production is of course the natural evolution of a business model without limits on profit and sometimes even guaranteeing profits by displacing social and environmental costs. Some have rightly argued that the problem is buried deeper within productivism—Western culture's reverence for production, reflected in the way we see nature as something that needs conquering with hard work, or in the way we see leisure as a source of guilt.[54] But in structural terms, society's utter obsession with energy expansion is also fueled by social architecture designed by the people selling us energy, who have us convinced that cheap energy is our right and that we should de-carbonize while carrying on normally. Just look at their take on energy efficiency.

Consider that Western societies tax cigarettes, alcohol, and sugary drinks to reduce their use and to fund interventions that reduce use even further, measures that directly conflict with producers. But electricity is different. Most U.S. states that legislate efficiency delegate responsibility to utilities and allow them to continue externalizing the various social, environmental, and ecological costs of electricity production, like air pollution and climate change. This is tantamount to entrusting Marlboro to set annual smoking cessation targets. Most electricity providers wish to maintain their typical business models in perpetuity, selling large quantities. Therefore, utilities have helped design efficiency programs that set modest benchmarks by reducing the electricity draw of specific fixtures and appliances. For example, ratepayers usually receive rebates for trading in their lightbulbs, refrigerators, and clothes dryers for more efficient models. Most utilities have enjoyed overall increases in electricity consumption resulting from Americans' increasingly electrified, digitized, and databased lives, not to mention hotter summers that warrant more air-conditioning. In other words, the efficiency improvements realized by rebates for efficient appliances come in tandem with overall increases in electricity demand, sustained in large part by the low cost of electricity.

Vermont's efficiency programs in particular are a model of success, as they have halted projected increases in future electricity demand, but, as in most states, in order to benefit from those programs homeowners must patronize approved vendors and then apply for rebates from relatively expensive retail prices. In Northern New England, where thrift and frugality are hardwired into local culture and identity, the corporate nature of "efficiency" programs is not lost on the public. One Vermont commenter writes: "I would much rather see my EV fee (*tax) go towards helping low income folks heat their home then towards a subsidized $20 LED light bulb made in China. . . . As any local carpenter will tell you, you can weatherize your home with local contractors for 25–50% less than you can with the Efficiency Vermont monopoly . . . who are only subcontracting out the work anyway." As the commenter points out, limiting winterization programs to formal business partners does not seem like smart climate policy. Nonprofit organizations targeting poor and elderly home owners have shown that weatherization can be achieved far more cheaply than what state-approved efficiency companies charge. It can reduce dependence on oil more than a lightbulb made in China, too. Thus, it seems clear that some electricity customers find fault less with electricity's high costs than with who profits from the high costs.

This is where the "pasta-throwing" strategy that opponents employ against wind turbines gets confusing. It is cringeworthy to hear the issue boiled down to one of subsidies and costs—that wind energy is bad because it costs too much. That argument comes from productivist regulatory logic that should be taking a back seat to climate logic—or humanist logic, as Ralph might argue. The modern fixation on electricity prices is a sneaky way of maintaining status quo electricity volumes because it reinforces consumer entitlement and keeps regulators on the task of ensuring that prices stay relatively low. The endgame for the low cost narrative is unequivocally more consumption, even amid talk of efficiency, because that narrative ensures that customers maintain their habits.

Not to mention, in the face of scrutiny in Northern New England, costs have proven a shaky argument for promoting wind because prices have exceeded forecasts, especially when accounting for additional transmission. Price increases drew negative attention to the subsidies, making wind turbines the target of fossil fuel interests and environmentalists alike.

When we attack subsidies for making electricity expensive or renewables cheap, we get distracted from the billions in global subsidies making fossil fuels artificially cheap and from their high human and environmental costs. We fail to see that the system is in need of redesigning because we are focusing on a tiny part of it. Subsidies are supposed to work by incentivizing costly investments in new technologies, so we should expect to pay a premium for technologies that have high potential for climate-mitigating benefit. When we critique policies for making wind development profitable or electricity more expensive, we reveal just how trapped we are in a reality shaped by energy coalitions. Low electricity prices would be fine if our energy came without consequences or victims, but we are very, very far from that reality.

Conclusion

Electric companies' successful lobbying to reduce solar feed-in tariffs illustrates how stuck in its ways the American energy system is. The problem cited by those companies—the disproportionate burden of grid maintenance paid by people who do not have solar panels—could be addressed by any number of policy alternatives if it were really the companies' greatest concern. They could charge uniform distribution fees for every grid attachment, as some states have, or limit payments to wholesale rates once solar owners recoup their costs; or legislatures could institute graduated price caps for low-income, low-usage households that might also incentivize conservation. Instead, electric companies, with the help of fossil fuel interests, have lobbied for policies that reduce the incentives for more homes to purchase solar systems, the opposite of what is needed to reduce fossil fuel dependence. Even Vermont has repeatedly dialed back net-metering rates, though it still allows large out-of-state solar developers to set up shop and export electrons, albeit under less favorable terms than before.

Rolling back feed-in tariffs is self-preservation for big utilities, as battery storage is becoming so affordable that any incentive to promote solar brings households one step closer to detaching from the grid, triggering the "utility death spiral." Large utilities claim to be champions of their lower-income customers, but these utilities enjoy enormous latitude to marshal evidence of their concerns behind opaque accounting practices defended by regulators. Given that wind energy benefits the world's richest institutional investors on the backs of ratepayers, it seems ludicrous that Northern New England utilities have targeted residential solar owners under the guise of equity. Obviously, industrial-scale wind turbines sync harmoniously with utilities' objectives, while paying premium prices for rooftop solar threatens their future. What is more, in lobbying for lower net-metering rates, utilities are engaging in a timeless PR war with the best ammunition there is: blaming the state for high prices.

It is not hard to see through this charade. Consider for example the "beat the peak" campaigns with which utilities use Tesla batteries in their customers' homes to minimize the percentage of the region's peak electricity load demanded from the grid. Utilities sometimes fail to disclose that reducing their overall peak demand reduces their share of responsibility for transmission infrastructure. This means that the relatively affluent customers of Green Mountain Power and Burlington Electric "beating the peak" are shifting the costs of maintaining the electric grid to ratepayers of other utilities, including small municipal utilities and rural electric cooperatives. Meredith Angwin writes, "The utilities are urging conservation in summer because they are playing the Game of Peaks. It's a utility game about money. If they play the Game of Peaks well, they can shift some costs from themselves over to neighboring utilities. Yeah, it's a zero-sum game ('I win' can only happen if 'you lose')."[55] These obscure politics of distribution again help reveal the duplicity in utilities' newfound concern with net metering and social equity. Utilities champion the cause of equity when it suits their bottom line but are happy overlook the inequity of displacing costs to customers of other utilities. Further, Angwin observes that utilities make no similar efforts to curb peak electricity demand in winter, when the climate impacts of New England electricity use are far worse due to more gas, oil, and coal use.

What is more egregious, a group claiming to be New England "ratepayers" has petitioned the Federal Energy Regulatory Commission to entirely usurp net-metering authority from states; but this "ratepayers" group has just fifteen members, who pay $5,000–$10,000 each in annual dues, and several members are current or former energy industry employees. For every statement supporting the petition, many following industry talking points, there have been approximately 1,000 dissenting statements.[56] It is likely that the fifteen members are technically ratepayers, but their links to utilities seem far tighter than their claim to representing all ratepayers' interests.

Why do regulators stand for this? In Northern New England, we can see that energy transition is introducing new pressures on regulators. Bureaucrats benefit more from having robust electric utilities in their state than from having energy-independent constituents. If middle-class rural electricity customers leave the grid, then utilities lose political and financial relevance. They will have less money for lobbying and for contributing to campaigns, and fewer jobs to offer former elected officials, weakening utilities' sway with policy makers. When policymakers' grant utilities lower feed-in tariffs, they may have other things on their mind than poor households paying for the grid.

Ridgeline wind has redrawn power dynamics in a more discreet, industry-friendly way than residential solar electricity. Adding turbines to remote ridgelines dispersed sites of electricity generation to places where the grid was poorly equipped to accept it, requiring expensive retrofits and major construction projects. Politics were retrofitted too. Green energy coalitions found themselves in a position to broker large and popular development projects. Mergers and takeovers in the region's energy companies brought access to capital and technical capacity that has made them major players in the projects. These consolidations have brought lucrative public-private partnerships to a region that has been devoid of major construction for several decades, lubricating the wheels of government, strengthening the collaborative culture of electricity coalitions, and weaving the region more tightly into urban energy economies.

The shock felt by so many citizens throughout the wind-development process has grown in part from having to see state governments with a history of accessibility and transparency align with the energy industry behind closed doors. State bureaucrats advanced energy agendas that promised swift and enduring benefits for large interests, many from out of state. Campaign coffers have benefited, but so too have careers, thanks to a revolving door between government and industry. Thus, among the reasons that so many have turned against wind is the perception that turbines are tethered to institutions that put profit before climate, and in the process, coopt elected officials and regulatory bodies. Moreover, this neoliberalizing of policies and politicians has unfolded under both regulated and deregulated market conditions, and with both Republicans and Democrats at the helm, testifying to the pervasiveness of the electricity industry's ideologies and associated strategies.

While Northern New England's progressive policy landscape created a welcoming environment for the renewables industry to collaborate with lawmakers, I have tried to complicate what people describe as "corruption." Rather than attributing shady dealings to individual actors and transactions, it is more accurate to characterize energy coalitions as a shady but legal system fostering a shady but legal culture of collaboration. I think Ralph, more than anyone else I have spoken to, understands this. On the issue of corruption, he told me: "It's not $20 bills behind the statehouse late at night. It's in the context of expanding these opportunities for Green Mountain Power and others to go at ridgelines and get $48 million. It's 'Hey David, we can do this, this, and this, and it will get you more money, and I'll be able to claim much more wind energy development and reducing carbon.'" Ralph knows that although this collaboration is not breaking any laws, it is shaping an energy system for the wrong reasons, resulting in a system with the wrong shape. The political economy lens helps clarify that this is less a product of conspiracy than it is a systemic and structured alignment of interests within the realm of neoliberal environmental policy.

My intention here is not to critique Northern New England's green energy industry or the region's many lawmakers committed to advancing climate change solutions. Legislators of small, rural states face enormous financial barriers. Climate change is a global problem. So should it matter that it is powerful corporations that build wind turbines? How else are poor states supposed to implement costly ecological reforms? Devoting sufficient resources to the task of reducing the need for heating oil and gasoline across the region would entail major shifts of policy. Promoting wind turbines offered a feasible way for legislators to affirm their commitment to the region's ecological ideals. Leaders did not conspire to create the culture of American politics that privileges financial largesse. If society requires carbon-friendly energy, then the entities crowding the political process and reinforcing the fossil fuel status quo must be challenged, and lawmakers must be presented with viable alternatives.

Nonetheless, regulators and policy makers have a part to play in remedying the antiwind rancor that they helped create. They should answer for the glaring shortage of data and accounting needed to validate that wind turbines are achieving the CO_2 reduction goals they promised. Regulators and policy makers should have to account for the limited oversight and enforcement deployed to hold developers accountable for their social and ecological operating mandates. And regulators and policy makers should account for weaknesses in siting legislation and exclusion from meaningful energy planning that propel global antiwind movements.

If indeed ridgeline wind turbines are a sustainable and efficient mechanism for greenhouse gas reduction, the perceived collusion between state government and "Big Wind" has cost the planet valuable time by introducing enormous dissent. Consider Vermont. If we can attribute Governor Phil Scott's election, at least in part, to his anti-wind sentiment, then we must accept that industrial wind's zeal to install turbines on Vermont's mountains helped to install a governor with lackluster environmental policies. Headlines at the end of 2020 announced that on Scott's order, Vermont will be conspicuously absent from a regional multistate climate collaboration taxing fossil fuels to bolster public transportation—an initiative that would more effectively reduce Vermont's emissions than selling wind electricity to Massachusetts, precisely what many anti-wind protestors desired in the first place.

Seeing Northern New England's energy landscape as a political economy clarifies that although there may be no explicit corruption at any one node in the system, in its entirety the system is corrupt, reiterating what historians have long argued. Opportunity structures, alliances, and collusion are essential to the functioning of our electricity grid because the American electricity sector has always been a public-private bureaucracy simultaneously tasked with defending profits and delivering cheap energy. Can a system with these conjoint objectives easily pivot to be a democratic sphere that responds effectively to global environmental crisis? I will argue through the rest of this book that such a pivot will require an uncomfortable reassessment of our renewable energy transition. We must see opposition to wind and the political disillusionment to grow from it as symptoms of an antiquated energy system. The following chapters explore how this antiquated system privileges fossil fuel interests and endows them with the cultural influence to design an energy transition of their preferred shape, proceeding at their preferred pace.

6

Scripted in Chaos

• •

In the spring of 2013, a colleague at Bowdoin College sponsored a screening of the energy film *Switch*.[1] The event was well attended, and the audience included several students from my environmental sociology course. The film begins with an awe-inspiring look at Norway's national renewable energy infrastructure, showing the inner workings of hydroelectric turbines buried deep within mountains to harness the gravity of glacier-fed alpine lakes. Oddly, *Switch* concludes with footage of an affluent white family riding a golf cart through suburban Texas, prescribing an energy transition that includes nuclear power and fossil fuels—to facilitate our move to renewables.

Discussing the film with my students the next day conjured plentiful criticism, without the slightest probing on my part. My students observed that the filmmakers seemed firmly committed to the forms of energy known to be severely damaging the environment, perhaps advocating for the energy transition we were on, but not the one we needed. It seemed incomprehensible that pairing renewables with fossil fuels did much to address myriad other social and ecological problems with our energy system or that climate change had even been considered. This tack is unsurprising given the film's narrator and content adviser is an economic geologist for the state of Texas with a decades-long career in oil and gas research, and given that the film was funded by and a distributed through professional associations for American geologists, including the American Association of Petroleum Geologists.

The film, however, and others like it, struck at an opportune time. Risk theory suggests that wind opposition reveals how shifts away from carbon-intensive energy are slowed because different stakeholders advance competing scientific analyses, and that as a result, the public's knowledge about energy is fractured across competing narratives. Amid this confusion, *Switch* offered the public a decisive transition road map presented on a silver platter by the energy industry; our fossil fuels are a trusty, reliable bridge to the magnificent green energy technologies awaiting us in our future.

Because many of us are committed to addressing the climate crisis, we rarely stop to interrogate who draws the energy transition road map, much less whether it sufficiently addresses the problems associated with extracting and burning fossil fuels. We would rather focus on accelerating the transition than criticizing it. This chapter dispenses with that undeserved deference, arguing not that our transition is technically flawed, but that it is orchestrated with little regard for climate change and thus far slower than it should be. It contends that the popular rallying cry for "energy transition," in which wind turbines are a central component, is in practice using the cover of neoliberalism to advocate for hybrid fossil fuel electricity systems while obscuring fossil fuels' role in the process.

Transition Chaos

Theories of energy transition suggest that fossil fuel interests are trying hard to steer the boat, even though they would rather it remain docked, because they have few alternatives and a pressing need for damage control. Sociologist Bruce Podobnik argues in his book *Global Energy Shifts* that energy regimes extend to every aspect of governance and civic life, and that their rise and fall occur in patterns.[2] Energy stability typically derives from and engenders political continuity and market dominance. Shifts from one energy regime to another usually entail the breakdown of hegemonic power and an increase in competition between states and firms. Additionally, periods of the most fundamental transition reflect chaotic circumstances characterized by sudden shocks and slippages.

Indeed, the current state of energy is well characterized as "chaotic," and long-established energy companies are feeling the effects. Acknowledging fossil fuels' limits and costs, the threat of climate change, and the acceleration of renewable technologies has created cracks in America's petro-electricity complex. Steadily, a more diverse array of businesses from a more diverse array of nations have entered the renewable energy sphere. Environmental sociologist David Hess sees corporate damage control as fundamental to modern energy politics, writing, "in the case of the green-energy transition in twenty-first century U.S., the political contestation by the incumbent industrial regime is so well organized that it should be at the center of the analytical framework."[3] I contend that where political contestation is futile, such as in progressive regions like Northern New England, the incumbent energy players' posturing and propaganda constitute attempts to manage the chaos overtaking their energy landscape.

Four characteristics are making the transition to renewable electricity uniquely chaotic. First and foremost is society's embrace of evidence that burning fossil fuel alters the climate. Even in the face of corporate suppression and denial campaigns, society's collective knowledge of climate change has spawned powerful green business and green finance sectors. These emerging sectors claim moral and ecological superiority over the fossil fuel sector, as well as financial superiority in electricity markets thanks to long-term price stability—assuming that the earth's sun, wind, and water are not drastically altered. These claims of superiority create a powerful public dichotomy between clean and dirty energy.[4] In the era of Teslas, cheap solar panels, and viable renewable-based grids and microgrids, petroleum companies are staring down a public relations nightmare, spurring some energy executives and investment fund managers to suddenly, and of course very publicly, change their ways—or at least pretend to.[5]

Second, the American oil industry is in the throes of unpredictable economic forces, making investments in renewables a potentially life-saving form of hedging. Fossil fuel prices fell to less than half of what they were before the global economic downturn of 2006–2009, but a majority of America's oil supply lay in geological forms that are expensive to tap.[6] Most surprising for fossil fuel companies, however, are emerging forces suppressing oil prices. For example, American driving habits are changing as more of us choose to live in cities and fewer young people learn to drive. Battery technology and all-electric vehicles are disrupting the market for automobiles with enticing design and quickly improving range. In rich countries, the COVID-19 pandemic has forced major societal changes, such as telecommuting and teleconferencing, which reduce energy consumption and just might stick around after life finds a new normal. More important, predictions for the developing world's explosive demand for fossil fuels are now tempered by the sudden onslaught of renewable electricity in lesser developed nations. China's expansion of solar and wind sectors have positioned it to dominate renewable energy manufacturing, and the impacts will reverberate across the world economy. Just as cellphones negated the need for much of the developed world to rely on telephone lines, renewables-driven microgrids could allow the developing world's rural poor to enjoy the modern convenience of reliable electricity without having to wait for enormous investments in centralized infrastructure, much less relying on combustion.

Third, unlike earlier energy transitions, the current transition has the potential to shift not just how energy makes people money, but for whom it makes money and on what terms. Current forms of renewable energy cannot be stored or transported like a railcar of coal or a tanker of oil.[7] Renewable energy's sources—the sun, wind, and tides—cannot be purchased and managed. The infrastructure for harvesting solar energy at the residential scale is mass-produced and affordable for a growing number of consumers, many of whom are willing to pay extra if it helps the environment. Innovative business models are emerging to harness income from this changing energy landscape, and many are new players using new skills.[8] To use a sporting metaphor, we now have a corporate energy field on which new teams engage in a sport that looks less and less like anything we have seen before—as if the National Football League were replaced with a football-soccer hybrid in which robots play for several individual millionaires instead of two teams of humans fielded by billionaires.

Fourth, characteristic of the current energy transition is ambiguity around who actually owns these robots, because our "clean" and "dirty" energy sectors are hardly as distinct as they seem. Marketing and public relations campaigns make it easier than ever to conceal a business-as-usual fossil fuel playbook behind greenwashed imagery. (For the best examples, watch British Petroleum's "Beyond Petroleum" ads of the past decade.) In fact, regulations designed to curtail carbon emissions promote dirty companies to engage in clean activities. Such are the outcomes of the European Union's emissions trading scheme, which allows European firms to satisfy local requirements by investing in renewable energy developments in less developed nations. These arrangements are shrouded from public view by neoliberal financial regimes. More than ever, capital flows freely across borders and is pooled and brokered through various institutional arrangements. As laid bare in Northern New England's experience with wind development, energy firms maintain diverse global investment portfolios through formal partnerships and nested ownership arrangements that facilitate collusion across firms responsible for generation, infrastructure, transmission, and sales.[9] Put simply, buying clean energy often supports companies still producing dirty energy.

In sum, this is the Wild West for electricity, as chaos exists at every turn. Ecologically reflexive thinkers, including consumers, politicians, and executives, wish to reduce society's fossil fuel dependence in order to, among other things, save the planet's ability to sustain human life. New capitalist actors and business models have emerged to take advantage of the unique profit opportunities in energy sources that are diffuse and hard to own. Major players in the fossil fuel domain, several of which have tried for decades to bury evidence of climate change, have adapted to renewables by owning them, in efforts to diversify and complement these companies' investments in extraction.

"It's the Gas, Gas, Gas"

From the chaos just described an image of coherence and order has emerged, thanks to a solution messaged with tactical precision by the American natural gas industry. In fifteen years, our national energy dialogue has shifted from debate about petroleum scarcity, clean coal, and carbon capture, to a more harmonious convergence around society's transition to renewables, just as *Switch* was preaching in 2013. "Transition" discourse is now common among fossil fuel interests because a gas-renewables electricity mix anchors petroleum firmly within the electricity sector, a sector made volatile in many states by climate legislation. An expanding footprint of American natural gas discoveries has lowered costs and unleashed zealous branding of a "clean, green" fossil fuel. Natural gas extracted by hydraulic fracturing, "fracking" for short, is even termed "unconventional gas," which manages to downplay the ecologically and economically costly way it is drilled.

Indeed, the modern renewables industry and the American gas boom have developed simultaneously and harmoniously. The American electricity grid has added intermittent wind and solar capacity with little storage potential, leaving the door wide open for gas to offer backup. This symbiotic development was no coincidence. In his book *The Boom*, Russell Gold describes how natural gas billionaire Aubrey McClendon aligned with the Sierra Club's national director Carl Pope to position natural gas as integral to a more climate-friendly energy grid, even as local chapters of the Sierra Club were resisting local fracking projects. In fact, several nationally prominent environmental groups once rallied behind the gas industry, helping solidify natural gas's status as a "bridge fuel" that would displace coal and facilitate the expansion of wind and solar.[10] The same renewable-gas narrative has been promoted by major fossil fuel companies in other nations with strong climate mandates, like Norway and Finland.[11]

In the Northeast in particular, natural gas piped from the Marcellus Shale region offers an inexpensive and readily available baseload supply to complement the intermittent wind and sunshine. This has resulted in a mixed electricity portfolio with appeal for several industries. In 2020, New England's grid was powered by approximately 40 percent gas. The question raised by Northern New England's wind opponents is precisely the same question that my students raised after viewing *Switch*; is the energy transition outlined above the transition that the global community needs, or the transition mapped out for us by dominant industry forces?

On the surface, the gas-renewables transition seems to provide cleaner energy along several fronts, but fierce debate persists about its climate efficacy. Natural gas can be transported efficiently across land through pipelines; the United States controls an enormous supply; natural gas burns cleaner than coal and oil, and natural gas syncs well with wind and solar. But this assessment does not account for the externalities of fracking, such as impacts on ground and surface water, aquifers and seismic disturbances, or staggering uncertainty about incidental greenhouse gas emissions during the drilling process, which could impose an enormous hidden toll in terms of the climate.[12]

Examining historical natural gas emissions frozen in Arctic ice, research released in 2020 finds compelling evidence that current atmospheric levels of natural gas components like methane have no historical precedent or natural explanation.[13] These findings point the finger at the gas industry's trademark shoulder shrugging and deflection tactics when it comes to tracking and reporting the industry's emissions accurately. For example, despite the "clean, green" industry PR, satellite data analyzed after a blowout at an Ohio gas well revealed that in twenty days, the accident released more methane than annual releases from oil and gas industries in Norway, the Netherlands, and France combined.[14] Yet the industry is quick to point fingers at other sources of methane, such as animal agriculture, hydroelectric dams, and food waste—each of which warrants attention, but none of which has ballooned at the same tempo as fracked natural gas. Experts see unaccounted-for natural gas release as one of the reasons why global greenhouse gas emissions have continued to increase and global warming has accelerated, even after coal's precipitous decline and growing electrification of transport. In response, the gas industry touts new advances in drilling equipment and techniques, in attempts to distance itself from the tidal wave of data eroding the industry's credibility on past operations.

As the wind turbine debate has brought to light, relying on natural gas to take over when the wind stops blowing means that new wind turbines are only as useful for addressing climate change as their source of backup power allows. Many existing gas generators are burning small amounts of gas to provide seamless backup power when the wind stops or the sun retreats. For all of these reasons, and because of Renewable Energy Credit trading, the complexities of regional grids, and weather fluctuations, the amount of unburned carbon that can be attributed to any one wind turbine remains a mystery.

While accounting for ecological harm is complicated, accounting for the benefits of our electricity transition that accrue to fossil fuel companies is simple. Slowly weaning the world from carbon is a strategy that allows the petroleum industry to maximize the productivity of its sunk costs in the extraction, refinement, and distribution of fossil fuels. Supplementing electricity grids with wind and solar prolongs natural gas supplies, just as corn-based ethanol prolongs the supply of gasoline. As we have seen, supplementing with renewables also compels investment in complementary infrastructure, such as ethanol plants, fertilizer and plastics companies, and of course wind turbines. Texas oil magnate T. Boone Pickens made headlines as an early adopter of wind power. Pickens began pursuing investments in wind nearly twenty years ago, when the costs of natural gas began to oscillate. Though the price of natural gas fell and stunted Pickens's plans, Texas now stands as a model in wind turbine development, having built a grid to move electricity from its remote prairies to its growing cities.[15] More than ten thousand turbines satiate the Lone Star State's expanding thirst for electricity while complementing, not competing with, the state's petro identity. Predictably, after a few hot summers with the threat of blackout, the conversation in Texas has turned to building more natural gas power plants to back the state's new renewables. Even as uncertainty abounds about natural gas's climate impact and long-term financial viability, *USA Today* reported near the close of 2019 that 177 new natural gas power plants were planned nationally.[16]

This synergy is most strongly reflected in oil companies' subsidiaries and investment strategies, which have included various types of renewables for decades. Recall that Vermont's award-winning Green Mountain Power is owned by a Canadian gas company, 38.99 percent of which is owned by oil pipeline company Enbridge, often in the news for oil and gas pipeline disasters. This corporate entanglement explains why President Trump's pullout from the Paris climate accord drew criticism from former secretary of state Rex Tillerson, Exxon's former CEO. Exxon Mobil owns the largest share of North America's second most productive natural gas–producing region, the Permian Basin of Texas and New Mexico, and profits from its interests in mixing corn-based ethanol with gas.[17]

To summarize, North America's energy industry has begun organizing itself around a long, slow ramp-up of hybrid generation, characterized by the deployment of renewable resources at the pace that costs and policies dictate, rather than what climate predictions prescribe. The motivation for this approach is common business sense, as companies' investments in the extractive business model are too costly, and typically too profitable, to simply abandon. America's strong energy path-dependence has been referred to by scholars as "carbon lock-in," because it epitomizes how deep structural commitment to one technology makes pursuing alternatives cost-prohibitive.[18] By contrast, renewables companies that have extricated themselves from the petroleum path, such as Denmark's Orsted, have begun promoting a fast transition, calling for immediate advancements toward an all-renewable electricity grid.[19]

Gas Expansion in New England, or Gas Expansion in New England?

In the summer of 2018, the centrality of natural gas to Northern New England's renewables portfolio rose to prominence when the regional grid operator, ISO New England, published a plea to state governments to subsidize expansions in natural gas capacity. Exemplifying risk-wrangling discourse, a series of rebuttals and refutations ensued on the *VTDigger* website. It started with commentary by an attorney with the Conservation Law Foundation, defending the region's progress toward a sustainable renewables-based grid that negates the need for more gas: "The clean power and energy efficiency we are already buying will keep the lights on and our homes warm during harsh winters, guaranteeing year-round reliability and energy diversity, all at a low cost to consumers."[20]

Several commenters criticized the author for exaggerating wind and solar's contributions to the regional grid, observing that, on some days, those sources generate no energy. On several winter days, without expensive foreign natural gas shipped through Boston Harbor, the region's natural gas electricity plants—which on some days account for more than half of New England's electricity supply—cannot operate because of gas shortages. Shortages are a recurring seasonal problem, as the region's gas pipelines are a bottleneck during the winter because gas for residential heating is given priority over gas for electricity. In the aforementioned advocacy, ISO-New England is arguing to keep an inefficient gas power plant from closing and for building more and bigger gas pipelines. However, a more nefarious set of influences has entered the conversation about New England's winter gas and electricity shortages. An economic analysis completed in partnership with the Environmental Defense Fund raised concerns that two of the region's vertically integrated energy companies, with interests in both natural gas and electricity, were reserving pipeline capacity on cold winter days, abandoning their reserved capacity at the last minute, and then reaping the rewards of resultant electricity rate spikes from the reduction in gas supply.[21]

Regardless of how effective or ineffective renewables are on the region's grid, or whether or not natural gas capacity was being deliberately curtailed, what is certain is the influence wielded by the American natural gas industry in deftly asserting its importance to the region. Consider, for example, the neoliberal tangle of price setting at New England's grid manager, ISO-NE, described in chapter 5. Governed by electricity insiders including power plant owners, ISO-NE's auctions convey explicit advantages for gas power plants to serve as baseload and backup power for renewables, attributed to the flexibility that gas power plants offer compared with coal or nuclear plants. Power plants that bid into the day-ahead electricity auction but then lose out to wind in the real-time auction do not have to repay their day-ahead payments if the price drops, or even if their electrons are no longer needed. Similarly, if a power plant has an accepted bid in the real-time market but loses out to a sudden glut of wind, then it need only repay ISO-NE at the new clearing price, which is typically reduced by wind's low costs, and it also saves money on the fuel it did not burn. In simple terms, these pricing policies are a subsidy for primarily gas plants to simply exist on the grid so that renewables can exist on the grid. [22] In 2021, the governor of Connecticut has argued that ISO-NE and its corporate alter ego NEPOOL are favoring gas even more explicitly by blocking renewable electricity from bidding into grid auctions at its true low price, which the state has subsidized in line with its climate goals.[23]

Gas power plants are so important to maintaining the current electricity grid that the grid itself can be cited as a barrier to a full transition from fossil fuels. Hypothesizing the complete removal of petroleum from American electricity, energy scholar Meredith Angwin writes, "If only variable renewables and storage were available, generation and storage-installed capacity would have to be five to eight times the peak-systems demand. Such a system would need reserve margins of 400 percent to 700 percent of peak demand. In contrast, on our current national grid, reserve margins of around 15 percent of peak demand are common."[24] In other words, by providing regions like New England back up for renewables, gas plants can be celebrated for saving us from billions of dollars of investments in renewables.

What is of particular interest to me as an environmental sociologist is the extent to which these explicitly pro-gas policies percolate into public debates in which natural gas is panacea for New England electricity. For renewables optimists, natural gas is a valuable complement to compensate for solar and wind intermittency. For renewables pessimists, natural gas is an essential substitute for intermittent generation sources with minimal value. Thus, the region finds itself caught between two stories of transition, both of which rely on natural gas. And to further illustrate how salient the natural gas industry is to the transition discourse, consider that the threat of winter shortages and rolling blackouts has not spawned major efforts to promote gas conservation. Instead of reducing households' vulnerability to winter price spikes through pooled heating systems or residential retrofits, the solution presented is simply to increase natural gas capacity, using the threat of blackouts to strike fear among the public.

Power Brokers in the Power Industry

Energy companies assert their transition vision through the policy partnerships described in chapter 5, but we must go a step farther to make sense of how that vision becomes resonant across society at large. It starts with politics. The ways that businesses convert their political power to broad policy narratives is central to understanding wind energy in Northern New England and the slow-moving national transition to renewables. Citing Levy and Newell, energy scholar Frank Geels contends that an "alliance between policymakers and business can turn into a stable and hegemonic 'historical bloc' if it also achieves consensual legitimacy in civil society via widely accepted discourses."[25] In other words, when interest groups outside the domain of business and the state adopt the concepts and languages advanced in the private sector, the private sector has achieved a formidable capacity to control public energy policy. Thus, we might consider the clear alignment between energy companies and mainstream environmental groups on both the issue of ridgeline wind and the language used to frame the issue as evidence of corporate hegemony in our energy transition. Chapter 7 explores how this hegemony extends beyond simply shaping energy transition to shaping the very energy reality we live in.

Evidence of transition hegemony can be seen in the priority given to cost stability instead of climate sensibility in the electricity dialogue in all three Northern New England states. Each state embarked upon its wind initiatives when the price of oil had reached record highs, and energy price uncertainty formed a central thread in the political rhetoric these states' energy coalitions deployed. As that discourse unfolded in the region's newspapers, the environment barely played a supporting role. Instead, wind turbines were framed foremost as intelligent, self-sustaining energy decisions given the volatility of global oil markets and hydropower from Quebec.

It is by fostering an insular political culture—boxing out citizen input— that coalitions advance one particular energy paradigm and exclude others. Geels describes how energy's status as a technocratic enterprise means citizens are shut out from shaping their energy systems because engineer-experts collaborate best with bureaucrats, who are accustomed to "a paternalistic and technocratic style of decision-making dominated by techno-economic arguments and cost-benefit analyses; informal consultation networks that give industry groups privileged positions as providers of technical knowledge and advice, with limited access for outsiders to close-knit policy networks."[26]

Geels writes extensively about the discursive patterns emerging in the United Kingdom's energy transition. Over the course of five years he traces the first appearance of renewable energy as a stand-alone option to its gradual framing as complementary to "large-scale technical options, which fit relatively well with the practices and interests of utilities and national governments. Other potential transition pathways are side-lined or marginalized on policy agendas (both in terms of attention and funding)." For example, Stirling observed that in the United Kingdom, the government embarked on a pro-nuclear publicity campaign, with its chief scientific adviser titling a public editorial, "Why We Have No Alternative to Nuclear Power."[27] Stated simply, while claiming to follow neutral expert technical advice, governments often support the particular low-carbon options that fit their industry collaborators.

Geels contends that a policy environment built around private input makes legislators more inclined to confer and collaborate with firms and experts than with citizens, municipalities, and other groups who engage in energy planning. Geels identifies this as a "post-political" technocratic style of governance that "favours existing regimes and makes it difficult to open up choices for wider political and cultural debate, which again helps explain why alternative transition pathways are side-lined."[28]

Geels reminds us that our energy transition is not a systematized or organized engineering project marrying the goals of sustainability and cost. Rather, the transition is a hodgepodge of profitable public and private initiatives made cohesive by shared narratives and vocabularies. In the United States, the private-public coalition of industrial wind supporters has shaped a public narrative, in concert with natural gas interests, in which wind turbines are an ideal form of renewable energy. This milieu helps explain why Vermont, with a single large monopoly investor-owned utility and a history of strong collaboration between lawmakers and business, passed two maligned energy bills that look nothing like those in surrounding states; the transition narrative is advanced by government collaboration with but one corporate partner. Geels observes that this type of public-private coalition is not resulting in energy policies engineered for climate change because the state is washing its hands of technological due diligence, leaving the technical details to the state's favored experts in industry.

Slow Transition Justice

Oil and gas interests have been so adept at shaping the energy transition that their impact can be seen in North American dialogues about the transition's fairness for workers, which support a measured pace for abandoning fossil fuels. As energy is a daily necessity for the entire world, "energy poverty" and "energy security" have become prominent dimensions of justice research related to energy transitions.[29] For example, price spikes or supply interruptions attributed to renewables will have more severe impacts on poor communities who may lose their service and be unable to afford personal generators or battery systems. But as some scholars articulate the need for energy systems without victims globally,[30] a parallel dialogue appropriates the term "just transition" to advocate solely for fossil fuel workers in rich countries. This approach advocates for shielding workers from catastrophic economic shocks resulting from retrenchment in fossil fuel industries.[31]

Extraction-based communities everywhere carry a burden for the global economy because corporations acquire land and mineral rights through dubious means, disrupt ecosystems, and sometimes sicken local populations. Workers are also captive to boom-bust swings and exposed to numerous dangers in their homes and workplaces.

Nonetheless, fossil fuel–dependent communities in the poorest countries have paid higher social and ecological costs than North American petroleum workers. State-sanctioned violence has murdered peasant and indigenous peoples in the name of oil in poor countries, and the industry has resulted in very few technical and engineering jobs for local citizens, reflected in the very small number of countries for which oil wealth has translated into broad middle classes or improvements in well-being.[32] What is more, in rich and poor countries alike, there are far more casualties of fossil fuels than just those at risk of losing their jobs; Neighbors to mines, wells, refineries, pipelines, power plants, gas stations, and waste disposal sites, as well as inner-city residents breathing toxic air, are all stakeholders in the energy transition too. How does a first-world "just transitions" narrative account for the long-overlooked rights and needs of these groups?

The co-optation of "just transition" as a first-world concept is remarkably convenient for the oil and gas industry. Of all industries to face an existential threat, those that extract a finite resource are the least surprising cases, especially compared with factory closures after trade agreements that promote off-shoring jobs, for example. How can oil, gas, and coal workers be surprised that their jobs are in peril, to say nothing of their bosses? If ever there were an industry that should be held accountable for the welfare of its own displaced workers, it is the industry that has poured money into lobbying, skirting regulation, and suppressing evidence of its sinister consequences for decades.

When focusing on fossil fuel jobs, "just transitions" reads like pandering to voters in red states to get on board with the green jobs agenda, which is ironic because paternalism risks further alienating displaced rural workers, and it is easy for promises of green jobs to ring hollow. As rhetoric in renewable energy discussions, "just transitions" plays like gas lighting, a conversation-stopper that perverts narratives of global justice and justifies a slow energy transition instead of the swift pivot the climate warrants.

Environmental sociologist Stephanie Malin has found that large energy interests benefit from a discourse of local victimhood because it reinforces the neoliberal individual as both the victim and the agent of reparation, not communities or the environment. Malin has studied communities polluted by fracking and others by uranium mining, where she observes that meaningful environmental justice takes a back seat to the politics of division between industry victims and industry beneficiaries. She writes, "I suspect that as market-based logic and individualized rights-based discourses become ever more hegemonic in the United Sates, questions of environmental justice will increasingly center on individualized rights to control land use decisions, whether or not those decisions are ecologically progressive or socially sustainable."[33] In other words, prioritizing individual remedies does nothing to guarantee positive environmental outcomes, especially where energy victims are desperate for recourse. Similarly, the "just transitions" narrative merely privileges a particular group of individual stakeholders in a way that insulates fossil fuel interests from pressure to address the larger, more diffuse trails of victims left in their wake.

Conclusion

Whether or not one finds fault with Northern New England's ridgeline wind, one can be certain that the region's leaders have tried in good faith to promote local renewable electricity. Maine, New Hampshire, and Vermont have each enacted rigorous environmental protections in other domains, and the public record along with my own conversations outline lawmakers' genuine concern for global climate change. How, then, do we make sense of the dissonance between their noble intentions and their support for a transition motivated by cost but not CO_2 reductions?

We must view American renewable energy development in the context of industry chaos, and deconstruct and decode the dominant discourse of energy transition accordingly. Regardless of the ideology behind legislation, ridgeline wind turbines play a role in sustaining a hegemonic energy agenda intended to withstand chaos. As energy businesses have accepted and to varying degrees embraced the need for an energy transition, their transition road map has gained widespread popular and political acceptance as legitimate, feasible, and agreeable. The road map feeds a public narrative of commonsense practicality and entails only slow modification of existing practices. Even the Trump administration has told Americans that a blend of natural gas, wind, solar, and hydro will power us through the next several decades, until the next major energy advancement unfolds. And this slow transition is sold to scholars and planners as the fair way to terminate fossil fuel jobs with no consideration of fairness for anyone else.

Slow weaning is not a new theme in the global energy transition. Adding wind turbines to New England's electricity grid is in some ways similar to mixing ethanol in our fuel. Richard Manning's 2004 essay "The Oil We Eat" paints a portrait of industrial agriculture's devastating effects on soil and dependence on fossil fuels.[34] Writing as Congress was increasing ethanol mandates, Manning explained eloquently why adding more corn alcohol to our gasoline would be a boon to the petrochemical complex already underwriting agriculture by fueling tractors and trucks and producing chemical fertilizers, pesticides, and herbicides. More than a decade later, we have seen that federal renewable credit programs governing corn ethanol's mixture with gasoline have also proven profitable to major oil companies and opened new opportunities for Wall Street investors betting on those credits' future value. Though we hear a lot about how the American government's ethanol mandate is unpopular with oil interests, the reality is that many of the largest companies profit from it—an unsurprising outcome for a neoliberal policy.[35] And, of course, slowing the rate at which we consume petroleum from the ground prolongs the profitability of petroleum companies.

History has shown that fossil fuel firms dabble in alternative energy but prefer that their core product remains the primary supplier of global energy.[36] Accordingly, fossil fuel companies are undoubtedly engaged in campaigns to halt progress toward renewables across the world; but where renewables have been carved a niche in which to grow, fossil fuel interests have extended their operations and investments to maximize profits within that niche.[37] This entails profiteering from renewable energy credits, promoting natural gas baseload power, and even using cheap wind and solar electricity for petroleum extraction.

The electric utility takeovers and acquisitions seen in Northern New England demonstrate the long-term, chaos-controlling business calculations energy interests are making to both adapt to and assert control over the energy transition. As the national fleet of electric vehicles multiplies, so too will the demand for electricity. Hotter, longer summers will compel more residents to use more air-conditioning on more days. But as electricity use expands, fossil fuel use declines. With every new electric vehicle on the road, the electricity sector becomes a larger feeding frenzy for oil money. Because the electricity sector is structured to accommodate renewables growth only to the extent that the sector is legally required or financially inclined to do so, it will remain a conduit through which fossil fuel wealth passes under most scenarios. This high degree of interdependence helps explain why fossil fuel conglomerates are now investing in electric utilities, why the petro-state of Texas invested in wind development as a complement to its overstretched petroleum capacity, and why Exxon prodded Donald Trump to legislate carbon emissions more aggressively. Society's most powerful energy interests and its network of lenders appear united in their vision for a transition to renewable sources that proceeds at the pace and on the terms these interests set, even as the pace of climate change demands a far more drastic change in our systems of consumption.

Those who believe that any forward momentum with renewables will result in climate progress must consider the plausible scenarios in which the dominant transition strategy exacerbates climate change. For example, simply replacing combustion engines with hybrid drive systems, thanks to the Jevons paradox, could add more cars of all fuel types to our roadways if gas prices continue to decline, accelerating carbon emissions. Similarly, if our wind-gas model of electricity generation were to one day reduce the cost of electricity and fuel greater electricity use, we would risk burning through more natural gas more quickly than if we relied on heavily taxed gas-only electricity. Put simply, the modest gains we make from an "all of the above" energy policy may be lost to accelerating consumer appetite for air-conditioning, video screens, and animal-centered diets.

Environmental sociologist Jesse Goldstein's perspective on planetary improvement as it relates to energy technologies is useful for understanding why energy transition is fraught with influence from petroleum interests and, more specifically, the way wind turbines run interference for a gas-focused transition strategy. Goldstein likens clean energy technologies to a "lightning rod for environmental hopes and dreams,"[38] which wind turbine opponents are sure to find poignant, as actual lightning strikes are one of opponents' expressed fears about ridgeline wind projects. Goldstein observes: "Lightning rods are meant to deflect energy away from the underlying structures that they protect. In this case, it is the infrastructures of a fossil fueled industrial economy along with the ways of life, or petrocultures, that is being protected."[39] In other words, grid-tied wind turbines should be understood as extensions of the grid. They represent and underlie the social ills perpetrated by fossil fuel extraction and the wasteful consumer lifestyles that a century of cheap electricity promoted. While turbines may symbolize positive change for most of us, they are simultaneously declaring the primacy of continuity. Goldstein explains why banking on conventional clean technologies is problematic:

> No amount of solar power, wind, or geothermal energy will, in and of itself, reduce the consumption of already proven fossil fuel resources. . . . While there is no need to categorically reject clean technologies just because they only deliver incremental change, we may need to question whether we can afford—collectively, socially, environmentally—to proceed with this faith in planetary improvement, and with what amounts to an unspoken acceptance of the fossil-fueled economy, perpetually improved but never actually undone.[40]

The recurring centrist reaction to this type of critique proposes that we think of our current energy regime as a crucial first step on the path to 100 percent renewable energy. While such optimism may be tempting to embrace, it is a tautological distraction. Natural resources are finite, making it obvious that humanity will one day rely solely on renewable energy. Because resources are finite, it is also imperative that our energy sources be recyclable or reusable, neither of which American policy requires. That is correct: we have no laws that our renewable energy sources be renewable themselves.

The issue of central importance to Goldstein, and myself for that matter, is not whether our current energy transition is a good start or better than no transition at all. The issue is whether the transition has been designed with the level of attention that future humans deserve. There is strong scientific consensus about the rate at which the climate is changing; if anything, early estimates of that rate appear understated. In light of that mounting evidence, the energy transition that we need is far more rapid and more forward-thinking than the path that has been laid out for us by our public-private energy coalitions, distracting us with lightning rods like wind turbines.

7

Why We Follow the Slow Transition Road Map

• • • • • • • • • • • • • • • • • • • •

When I ask my respondent Christine what she thinks the best message for wind opponents to communicate to the public is, she thinks for a moment and responds, "It's challenging . . . everybody believes that wind is free. It's producing no pollution, and what's not to love about it? If you start to address, 'Do turbines reduce carbon?' then they change the channel to a different argument . . . that we have to get off fossil fuels." When Christine says they "change the channel," she is describing how most people cannot readily accept that a wind turbine could be anything less than an ideal solution for fossil fuel addiction. But wind resisters like Christine are very worried about quitting fossil fuels too. If everyone has a common goal, why can they not agree on a solution? Why is Christine's point, that there may be faster, greener ways to quit fossil fuels, such a hard sell?

I argue that the reason is that energy coalitions get a lot of help from us—from environmentalists, from the public, and from ratepayers, who accept and reproduce these coalitions' slow transition vision. Renewable energy research lacks an effective frame for probing not just why specific interests adopt specific policies, but also why the public is so willing to go along with those policies despite having limited knowledge about them. What sociology can lend to this gap is an approach to understanding how the control of knowledge takes on a powerful role in producing culture, the unspoken rules and practices molding our communities and ourselves. Widespread tacit support for a slow energy transition can be understood as the management and shaping of our energy knowledge by fossil fuel interests. This chapter outlines three ways that a slow energy transition relies on knowledge, or a lack thereof, among the American public: (1) an energy culture designed for passive consumers expecting low prices, (2) wind turbines' status as totemic icons of environmentalism, and (3) renewable energy's foreboding implications for equity in an unequal world. These three elements amount to strong barriers within our psyches and belief systems that hold us back from an equitable and sustainable energy transformation.

The Passive, Pliable Consumer

Driving the hybrid transition message, fossil fuel interests have created the reality in which supporting the renewable "energy transition" entails putting our support behind fracking and pipeline expansion. For energy historians, this result is unsurprising. With public relations expertise spanning nearly a century, electricity companies, often overlapping with the gas industry, have subtly outlined how the public understands and talks about the electricity grid's transition from fossil fuels, portraying it as a slow weening under the watchful eye of our utilities and their regulators.

Energy interests craft this portrayal by engineering the entire universe of energy sources and systems that we use, including our norms for electricity use and our vocabularies for talking about it. Put differently, the way the public understands electricity is the result of a specific recipe, developed and perfected over time by politicians and businesses, a manifestation of corporate influence in the legislative framing of electricity. Ideas become powerful when they reverberate across diverse groups who embrace them as truth. Energy scholar Frank Geels contends that energy companies consolidate power when they spawn environmental policy discourses and vocabularies that become internalized and accepted across audiences.[1]

This framing of energy companies' cultural sway resonates with how sociologists explain tactical deployments of power. Whereas the earliest sociologists understood power as tangible influence and coercive capacity, later scholars noted how power operates as a more diffuse social phenomenon. Twentieth century sociologist Michel Foucault discussed the particular power that emerges from language and dialogue. Discourse, as understood by Foucault, shapes knowledge and social realities so that everyday habits and beliefs are ingrained with power relations. Accordingly, discourses are both the architecture for creating and reifying what we perceive as reality and also the mechanisms through which actors capitalize on their social position. Discursive power, therefore, refers to the capacity for narratives—that is, talking points and informal scripts—to condition the social world in ways that benefit those who generate and sustain the narratives. And in this way, language serves as a medium through which the creation and distribution of knowledge becomes a tool for wielding power. Members of Northern New England's energy coalitions have embraced a mutually beneficial model of wind development in a way that drowns out counternarratives, an essential tactic of exerting their influence through shaping discourse.[2]

Discourse becomes powerful when it infiltrates our beliefs and practices. Before Foucault, Italian sociologist Antonio Gramsci observed that elite groups establish cultural norms that when adopted by social subordinates serve to benefit elite interests. He coined the term "cultural hegemony" to characterize instances of this hard-to-see social persuasion that lives ingrained in our daily, unquestioned habits and outlooks. Hegemony here connotes power that can be exercised without physical force and that is, to some degree, accepted by those it acts on. Sociologist John Gaventa's research in Appalachia applied this perspective to understanding political quiescence in communities oppressed by coal mining companies. Building from Steven Lukes's theory of power, Gaventa showed how local cultures incorporated norms and values asserted by coal companies' managerial class, giving way to the pervasive belief that what was good for the company was good for the community. Forty years later, it is common to associate this energy Stockholm syndrome with petroleum-dependent communities. In contemporary national politics, the media frequently probes why energy communities defend the very same corporations that treat energy workers as disposable and replaceable.

While it is easy to see the norms that subject coal miners to their employers, we pay less attention to culturally hegemonic norms that subject all of us as consumers to the whims of energy companies. We rarely question how the reality of our electricity sphere has been created by those who sell us electricity—how we expect electricity to be cheap and available at our homes, how we expect it to be ceaseless, how we passively consent to noncompetitive markets. We submit to powering virtually every aspect of our lives as individuals, buying an ever-growing shopping cart of electric things with short life spans, many of which could be replaced by shared or hand-powered alternatives. All of the subtle habits that make us consistently profitable ratepayers have been constructed over time, both by shrewd marketing and by the reifying ways our society understands and talks about electricity. Our energy habits perpetrate severe ecological harm, and yet our choices for fixing our predicament are few unless we can afford expensive alternatives. Within this electric universe, the slow-weening energy transition discourse is a reality that was slowly spoken into being by public-private energy alliances.

Energy Culture: How It Is Made

Technocratic private-public energy coalitions enjoy political authority in part by engineering powerful cultural threads that the coalitions weave into everyday life. Recent scholarship has argued that most energy consumers are beholden to a culture in which utility companies dictate rules and norms about how the public uses and thinks about electricity. As Patrick Devine-Wright observed in the United Kingdom, the norm in centralized energy systems is to treat customers as outsiders by maintaining strict boundaries of knowledge between energy insiders and the public. Customers are treated as simple, insatiable addicts, which leads to a barrier of knowledge and technical capacity that sustains the status quo by preventing disruptions that might maximize conservation and efficiency. Devine-Wright writes that "the centralized energy system is embedded within, and has helped produce, a social representation of the 'energy public' that is overwhelmingly characterized by deficits: of interest, knowledge, rationality and environmental and social responsibility. . . . This is a self-fulfilling prophesy."

To put these remarks into context in the United States, energy is of enormous consequence to every aspect of our existence. Electricity's availability, price, and consistency are characteristics that shape how we structure our days, and for those who fall asleep to the whirr of fans, white noise, or television, these qualities of electricity are essential for sleeping through the night too. More fundamentally, electricity's availability, price, and consistency shape our conception about what is normal, acceptable, and desirable about energy itself. We are conditioned to expect that a light switch will illuminate a room instantaneously, at minimal cost, and with no greater exertion than a finger flick. We rely increasingly on screens to keep our kids occupied, in homes, cars, and restaurants, and we expect hotel rooms, planes, and airports to be outfitted with abundant charging ports. We have built cultural institutions, including family rituals, treasured recipes, and modern vocabularies, predicated upon cheap energy, available on demand. This has all happened slowly over time under the watchful eyes of the companies that build our appliances and sell us the juice to power them.

The same watchful eyes that got us hooked on electricity are helping to herd us into an energy transition that swaps one fossil fuel for another, so that we continue to consume at high volumes. And thinking back to the industry-funded film *Switch*, it is clear that the institutions selling us our energy want very much for their vision of transition to seem like a natural and commonsense solution. For example, capitalizing on the academic credibility of its funders, *Switch*'s release entailed a well-publicized rollout targeting universities and community groups across the United States, which is why it played on my campus. It included a website and teaching curriculum, complete with discussion questions and worksheets for high school and college students. It was not enough for this film to simply be seen; its backers wanted the film to be seamlessly folded in to students' environmental studies education and to guide the conversations that resulted.

Switch follows in a long line of narrative-shaping content delivered by fossil fuel interests to educate viewers about energy through the eyes of energy companies. The same year, petroleum interests backed surreptitiously by Shell Oil released a documentary, titled *The Rational Middle*, which American colleges and universities also validated through sponsored showings. The year 2012 also saw the release of *Truthland*, a gas company–backed response to Oscar-nominated *Gasland*. Partnering with schools has become standard for petro public relations.

It goes well beyond low-key propaganda like public film showings. Scholars Benjamin Franta and Geoffrey Supran penned a 2017 article in the *Guardian*, titled "The Fossil Fuel Industry's Invisible Colonization of Academia," reporting that the majority of energy research centers at American universities are funded in large part by fossil fuel companies, including those at Harvard, Stanford, and MIT.[3] And apparently it is never too early to start spreading the petroleum gospel to American children. A 2017 report by National Public Radio highlights the Oklahoma Energy Resources Board, which has channeled forty million dollars into public school curricula to normalize, if not celebrate, petroleum extraction. In a state experiencing earthquakes from hydraulic fracturing, school children learn from a baby-faced, moonwalking, bowtie-wearing "Professor Leo" that fracking pads are "quiet and serene."[4]

Corporate messaging using psychological and cultural research has helped construct our energy consciousness for decades, so it is important to understand "transition" marketing like *Switch* in that historical arc.[5] The oil and natural gas industries have learned that any promotion of "transition" makes them appear more forward-thinking than they actually are. By advertising themselves as purveyors of the nation's "bridge" fuels, they lend progressive credibility to their business operations among Americans rightly concerned with the energy status quo. "Transition," after all, connotes progress. And if the message comes from domestic companies paying stalwart American workers high wages, then it gets the "all-energy-is-good-energy" constituency on board too.

To achieve the transition they want, fossil fuel companies are trying hard to project a forward-thinking, solutions-oriented image, to hide the filthy tracks of their past from public view. Author Malcolm Harris was invited to a strategic brainstorming retreat by Shell Oil's "Scenarios Team" in 2019 because he had written a groundbreaking book on millennials. He recalls the event in *New York Magazine*, quoting a senior Shell economist: "We're going to get as much out of [oil and gas] for as long as we can." And on their strategy for managing to do that as global climate change worsens, Harris writes, "denial-backed delay is no longer sufficient, it seems. They're now hoping to leverage their incumbency, and fossil-fuel wealth, to lay claim to the world's clean-energy future as well. To do that, they'll have to persuade young people to forget who caused climate change in the first place, or at least to let bygones be bygones." Harris suggests that fossil fuel companies are fine-tuning their PR to work on a younger, more critical generation, to convince younger people that fossil fuel companies are part of the new-wave, solutions-oriented, clean energy business world, even as they continue to prospect for oil.[6]

Energy companies, with their vast resources, enlist a lot of help in shaping reality. This is why Shell Oil wanted to learn from the man who wrote the book on millennials. The media and journalists are active partners in sustaining energy culture, and now more than ever it seems that they favor feel-good stories about energy transition over more critical treatment of energy issues, Malcolm Harris notwithstanding. In an empirical analysis of renewable energy journalism, environmental scholar Ozzie Zehner found that the media exhibit a strong inclination to promote energy-generating solutions instead of energy conservation solutions, implicating media ownership's interest in maintaining high levels of consumption.[7] Zehner contends that obsession with new ways of generating electricity amounts to "techno-denial" of the possibility that growth is unstainable, allowing one to believe instead that technology can solve all environmental problems.[8] These self-soothing tendencies of ours make it difficult to pierce the public relations armor of an upbeat transition narrative scripted by energy businesses.

Zehner himself became the target of green media scrutiny for trying to pierce this armor when he teamed up with Michael Moore to produce a controversial exposé of renewable energy, *Planet of the Humans*.[9] Critics took very vocal exception to the film when it was released on YouTube in early 2020, and I believe much of their criticism is justified, but they piled on Zehner in particular for claiming in the film that a specific concentrating solar electricity plant in California may use more greenhouse gases than it saves.

This claim, like most energy issues, is complicated to verify, but rather than try, several critics instead referenced energy life cycle research on the hypothetical performance of similar plants, even though Zehner based his statement in the film on one plant's operating record, not generic estimates.[10] This is made clear on the film's website.[11] One critic accused Zehner of being biased against all renewables and questioned his claims about mining rare earth minerals because, the critic alleged, these claims could not be corroborated with a Google search.[12] One would think that a published author and the co-producer of a film about energy used more investigative tools than just Google, and yet that was as much effort as the critic was willing to invest in refuting Zehner's argument.

Zehner's general point is that when renewable technologies are deployed in the real world there are hidden demands for fossil fuels propping them up, made worse for the climate by the fact that we are expanding our grid and using more electricity overall. That is, our transition is slowed both by its continued reliance on fossil fuels and our continued growth in total energy use. Zehner does not refute life cycle analyses or claim that technologies are incapable of being improved. He is pointing out that an existing technology is overshooting its life cycle analysis and advocating for meaningful and expansive efficiency. He is hardly the only, or even the first to advance this perspective.[13] As is typical in risk society, however, competing sides are using different science to convince their audiences that one side is right and the other is wrong. I cannot say who is right—just that they are not even debating the same point or using the same measurements. Uncertainty, opacity, and exclusion of the public make it easy to weaponize science this way. All of this is to say that the indignant criticism of Zehner that rolled in following *Planet of the Humans* seems to prove his point about media bias—that it is indeed very hard for the media to swallow critiques of their favored and fetishized renewable energy options.

Recent investigative journalism suggests that major news media have more than a general interest in promoting energy use. Energy coalitions and their scripts are the result of tactical psychology perfected by oil industry PR and advanced in America's premier newspapers. *Drilled*, a groundbreaking podcast exposing the energy industry's long history of strategic communications, details how Exxon Mobil targeted policy makers and business leaders with "advertorial" essays in the *New York Times*, with the expressed purpose of delivering its message to the nation's decision-making elites.[14] Doing so framed the parameters of energy policy around boundaries set through provocative and carefully worded industry narratives.

The tactics used by today's *New York Times* and *Washington Post* make the advertorial obsolete, as the newspapers now produce informational content on oil companies' behalf that blends in with news stories or film reviews. Of this technique, *Drilled*'s Amy Westervelt writes, "In 2020, influence doesn't look like an oil tycoon in a top hat showing up at your desk to twirl his mustache and tell you to spike a story. It looks like readers being fed a bunch of oil propaganda before, after, and right next to your legit climate reporting." It is hardly surprising, then, that the film *Switch* was celebrated across the national media landscape for its "rational" and "balanced" perspective, and for being "free from hot air." Fawning praise from the *New York Times* carried the headline "*Switch* Explores a World of Fuel Options,"[15] and the *LA Times* heralded "*Switch* Sticks to Energy Details."[16] This coverage, peppered with the language of optimistic futurism, goes a long way toward making the energy transition advanced by fossil fuel interests seem like the only viable path forward. The headlines imply that alternative paths are somehow impractical if they exclude fossil fuel options and that these alternatives are usually too focused on the big picture, with too few "details."

Most of us accept corporate messaging because we cannot fathom, or would rather not acknowledge, that corporate interests can be so far off from our own interests. Apart from raw political power garnered through public-private alliances, Australian researchers studying corporations and climate change contend that widespread belief in corporate legitimacy, predicated largely on our utter dependence as consumers and workers, empowers businesses with a supportive public narrative holding that "what is good for corporations is good for us all."[17] We often defend business, appreciating when it adapts to consumer demands, but we hate to acknowledge that businesses are adept at implanting and nurturing those demands. It may be for these reasons that many celebrate oil companies' slow greening as forward momentum in the climate fight instead of seeing their slowness as willful intransigence. That is a bit like rewarding a puppy for pooping on the floor instead of your shoes. It is progress, sure, but there are still giant messes for us to clean up, and will the puppy ever learn?

Electricity companies, which are often purveyors of gas as well, have gained the upper hand in messaging by utilizing a century of marketing and PR savvy and by amassing wealth over decades of monopoly power to ply the media. More important, however, is that status quo energy interests benefit from their positions as authorities and experts at the center of the only energy system their customers have ever known. No matter their sins, and they are many, electric utilities provide us a service on which we depend and that requires little engagement in return, a tremendous asset in uncertain and tumultuous times. It is a paternalist consumer relationship that affords us our Netflix, our dishwashers and clothes washers and dryers, and most of us are not inclined to devote precious time to spearheading the design of a more sustainable energy reality when it seems that it is already unrolling.

To summarize, our collective imagination about future electricity systems is limited by our experience of electricity in our daily lives, at the switch, the on button, and the utility bill. Exploiting our passivity, we are shut out from energy governance, which allows electric utility boards to boil our interests down to the single issue of price stability. Energy industries, meanwhile, bombard us with razor-sharp and carefully targeted PR to build our confidence in their technologies and general business outlooks. When we support an energy transition that simply replicates a centralized model of generation and distribution, we are acting within structures of an energy reality created and reinforced by those profiting from centralized generation and distribution. The language of transition deployed by the American energy sector is seeding a landscape that the industry plowed itself.

While scholars have focused extensively on greenwashing as an enterprise aimed at consumers, a dearth of research frames the energy industry's persuasive discourses and technical policy prescriptions in a similarly accessible way. Social scientists are, therefore, limited in our familiarity with the industry dog whistles used to rally the public around the "environmental" practices that prioritize energy shareholders at the expense of sensible climate policy. Picking apart the public narratives around renewable energy reveals that well-intentioned lawmakers and an environmentally progressive electorate are supporting an energy transition advanced by oil and gas companies: slow, steady, and inclusive of fossil fuels. This energy worldview amounts to much more than corporate greenwashing. It is creating a parallel reality in which the public recalibrates what it thinks the color green looks like.

This argument is a complement to environmental sociologist Andrew Szaz's finding that promoting eco-conscious purchasing lulls green consumers into a false sense of eco-empowerment.[18] My argument posits one additional nuance: that in the domain of electricity, our energy institutions use wind turbines to signal that we can help the environment *passively*, in some cases without doing anything extra. This is doubly deceiving in that not only are the climate benefits of wind electricity difficult to ascertain, but local wind turbines may actually worsen our personal climate impact if the RECs are sold off.

Some electricity companies are experimenting with giving consumers more energy knowledge, but the process has been fraught. Many American electric utilities have implemented demand-side pricing and education campaigns to incentivize conservation and efficiency by balancing their load across times of typically low and high electricity demand, so that high evening prices compel consumers to save their energy-intense tasks for cheaper hours. Spreading out demand can reduce reliance on the inefficient generators kept on standby for peak demand.

"Smart" meters, rolled out in roughly half of American households, facilitate this knowledge by showing ratepayers real-time costs and tracking their consumption habits. They prompt consumers to see their habits not as isolated individual decisions, but as linkages in a chain of relationships. In terms of the broader public discourse of transition, these sorts of programs represent monumental change because they unite our policy discussions about cost and about climate change instead of prioritizing one over the other. These policies are implemented to limit greenhouse gases first, and let the prices fall accordingly.

Deconstructing the Symbolism of Wind Turbines

While smart metering forces customers to engage actively with energy by feeding us information and tracking our usage, wind turbines are prominent visual symbols of society's energy transition that we can engage with passively, as we are accustomed to. And they are pretty. Their stoic form embodies many of the characteristics that rural New Englanders, if not most Americans, value. When seen from a distance, they raise no questions of size or scale. They appear tethered only to the breeze. They do not release soot or destroy rivers, and thus it seems that wind turbines hide no dirty secrets. They are symbols of local people "helping." But how much more do we know about wind turbines?

Totemic Icons of Environmentalism

Because they symbolize purity for so much of the public, wind turbines do not receive much deep thought. They are part of an energy system with many moving parts, but when we support wind electricity we think only of the part that twirls overhead. We assume the turbines produce energy transparently, by turning a generator with every revolution. What seems so simple and uniform belies great complexity that one sees only when looking closely. And even then, some questions do not have readily available answers. For example, it is unclear how much electricity a turbine draws from the grid, how much fossil fuel is needed to run peak generators, and thus how much carbon is offset by any given wind turbine at any given time and place.

As much as wind turbines might make us feel good for how we *think* they are helping, not all installations are the same, because not all electricity grids are the same, and neither are all project sites. As Christine has learned, most people cannot fathom that fossil fuels are used to back up wind turbines. And they do not think about carbon footprints accounting for deforestation, blasting, road building, and the transit and construction of the turbines themselves, or that projects might never meet their promised capacity because of grid curtailment, less wind than expected, or compliance measures to reduce noise.

Our human inclination to overlook complexity within the familiar imbues wind turbines with special status, which I refer to as totemic icon status. Christine kept confronting this status as she tried to educate the public about wind turbines' complexity. As she grew more resistant to wind development, it seemed that the public's zeal for it only strengthened. She shared with me a photoshopped image depicting wind turbines along the waterfront of Burlington, Vermont. The doctored photo had appeared on a website promoting a fictitious wind project. The fake turbines rose from the waters of Lake Champlain, silhouetted against its spectacular shoreline at sunset. Though the image was intended to rile NIMBY impulses in Burlington, public commenters expressed hope that the fake project was real. Christine lamented such superficial wind love. Those commenters had no idea what ecological sacrifice the construction of industrial wind turbines would mean for their waterfront.

While Christine was concerned with what those commenters did not understand about turbines, I find it useful to consider the origins and character of what the commenters thought they did understand about turbines. Turbines symbolize environmental progress and are universal icons of it. They started as powerful brand ambassadors for renewable energy, climate awareness, and the environmental movement more generally, like polar bears, the Toyota Prius, and the circular arrows on our recycling bins that can somehow make even the plastic industry appear green.[19] As Zehner describes more poetically, "In a process of self-fashioning, environmentalists offer their arms to the productivist tattoo artist to embroider wind, solar, and biofuels into the subcutaneous flesh of the movement. These novelties come to define what it *means* to be an environmentalist."[20]

But wind turbines have become more than sacred monuments for passionate environmentalists; they are now icons of a newer, cleaner energy system among the general public. "Icons" are symbols that convey meaning instantaneously, requiring no thought—a cognitive impulse that accommodates our hurried and preoccupied lives. Most of us will never observe wind turbines up close or hear them through our walls on a windy night, but they seem simple enough that we feel we understand how they operate. Many ordinary Americans see wind turbines as commonsense technology that is finally being deployed to generate local electricity free from pollution and foreign meddling. From coast to coast, wind turbines adorn postal service marks, coasters, postcards, truckers' hats, and children's cups, symbolizing energy sensibility in direct opposition to the symbolism behind gas pumps, oil refineries, and nuclear cooling towers.

As beacons of identity and place attachment, wind turbines have also taken on status as prominent totems in Northern New England. Totems are elements of the natural world whose image was adopted by indigenous cultures as emblems of their communities. In the tradition of the Pacific Northwest's Native peoples, totem poles are wooden edifices, often tall, depicting these representations in story-like fashion, with emphasis on conveying familial connection. To the beholder, totem poles represent place knowledge and belonging to the family, clan, or tribe. As I have observed in Northern New England, wind turbines convey to their beholders both a sense of individual environmental affinity and faith in the community to which they belong. That is, wind turbines foster a sense of belonging by broadcasting that one lives among others who hold similar beliefs. Throughout Northern New England, the turbines nurture a positive sense of self by affirming that one's personal environmental beliefs align with one's actions as an electricity consumer.

In Vermont in particular, communal eco-identity is reinforced in both subtle and not-so-subtle ways. Officially and by literal translation of the state's name, Vermont is the "Green Mountain" State. Its most iconic brand is Ben and Jerry's. The state has banned billboards since the 1970s. Its cars don luminous green license plates, but if that is too subtle, one can buy an oval bumper sticker at a convenience store that reads "Vermont: As Green as It Gets." The children's museum in Burlington features a tabletop interactive exhibit dedicated to wind, with tubes that shoot balls through the air and a small-scale barnyard wind turbine, the likes of which I have never seen anywhere else in the state. It struck me as peculiar that with the controversy over wind energy, wind was presented to the public with such simplicity, but that exhibit is a testament to how resonant the perceived purity of wind energy has become.

In Northern New England, pro-wind sentiment often reflects self-actualizing statements about what turbines represent to society and themselves. A common refrain at public meetings is "we're doing our part," as many citizens claim to be reminded when they see nearby turbine blades turning. By the numbers, support for renewable energy has been buoyed by residents of the region's most urban areas. It was residents of Vermont's Chittenden County who demanded a state carbon tax at the public forum

I attended in Essex Junction. These are high-earning progressive residents deliberately choosing to live in beautiful places, who are accustomed to broadcasting their environmental beliefs by wearing Patagonia clothing and driving Priuses and Teslas.

I suspect that many across the developed world derive pro-wind beliefs in part from a team pride that reinforces deeply personal aspects of their individual identities. Yes, they care about the environment and climate change, but being prominently visible in their fight for the planet is important too. This visibility provides a sense of team pride and personal affirmation by celebrating community accomplishments. It is why Vermonters welcomed a fictional rendering of turbines on Burlington's iconic lakefront. Burlington's municipal electric department is already a national leader in renewable energy, but it burns a lot of wood, buys its wind power from projects well outside the city, and engages in REC arbitrage, which means the city hosts no prominent local trophies testifying to its energy progressivism. But the city's citizens are green, and they are proud of it. They want the world to know they are green, and they like being reminded of how green they are. The city's inclination to support wind turbines along the region's ridgelines constitutes an eco-tribalism yearning for a totem.

By contrast, Northern New England's more rural citizens, many of whom dry clothes in their yards, grow and can their own food, and chop and stack their own wood, may derive less satisfaction from the symbolism of wind turbines. Their reverence for the natural world is not complicated by the adoption of an urban or suburban lifestyle. While both groups care deeply about the same things, one finds comfort from turbines' symbolism, and the other has problematized the symbolism.

Outside of New England, a different network of stakeholders promotes wind, and the mix of traits projected onto turbines' physical form in those places is different. Where wind is stronger and more consistent, such as the American plains, the purity narrative plays a subordinate role in the symbolism of wind turbines. In plains states, energy independence and the profits accrued by owners of large landholdings are sufficient to sell turbines as being locally desirable.[21] But turbines are a harder sell in places where coal, oil, and nuclear lobbying portray them as ineffective or dangerous. Thus, it seems unlikely that Texans, West Virginians, and Vermonters derive the same symbolic value from wind turbines. And as I discuss later, turbines' totemic iconic status may be precisely what hurts wind development in some places.

The Prius of the Grid

The problem with totemic icons of environmentalism are the mental invest-ments we make in knowing them without really knowing them. Wind turbines are complicated things. They are more than whirligigs, not only because of their mechanical complexity and the diversity of our grids, but because like any technology, their value to the world is derived through social interactions. Scholars of science and technology studies (STS) recog-nize technologies' social dimensions as foundational. More than a stand-alone apparatus, a technology embodies a socio-technical system of interests and actors invested in the technology's success.[22] Moreover, a technology is useful only insofar as humans agree that it is useful, and that agreement is the product of negotiations, some of which are public while others occur behind the scenes, in laboratories, boardrooms, and courtrooms.[23]

Modern wind turbines are a technology created and promoted with visions of ecological grandeur. They are championed under the assumption that they achieve the coveted "triple bottom line," securing profit, social jus-tice, and sustainability. But we must understand that this sales pitch is a narrative actively projected onto a physical form by those with a vested inter-est in promoting that form as a technological innovation: its designers, manufacturers, investors, and shareholders. Ultimately, wind turbines' prev-alence has far more to do with the resonance of that narrative than with the technology's accomplishment of greenhouse gas reduction in any uni-form or universal sense. This helps to explain why wind turbines prolifer-ate in New England, despite limited evidence of their ecological value. They embody a socio-technical system that is financially rewarding to a large net-work of promoters and benefits from a pervasive narrative of ecological purity. To be clear, this is not to suggest that as electricity sources turbines are worse than burning fossil fuels—just that they are not maximizing their potential on ridgelines in New England and that in places where turbines could be maximizing their potential, they are not utilized enough.

In these ways, wind turbines are a lot like the Toyota Prius. Environmen-talists have coded wind turbines deep in our psychological vocabulary as the green salve for our bad habits. Their shape is distinct and symbolic. In many instances they meet specific ecological goals, but like Priuses, wind turbines make different contributions to carbon reduction under differ-ent socio-technical circumstances. Though Priuses and other hybrid cars are prominent symbols of energy transition, CO_2 reduction varies across contexts and cannot rival that offered by mass transit. On a suburban highway commute, the climate case for hybrid cars is weak if solo drivers access the carpool lanes or avoid public transit. In other words, a hybrid car offers dubious ecological benefit if it's cruising at seventy-five miles per hour on the highway passing half-full trains and buses. Even in rural areas, where hybrid vehicles are one of the few options to reduce fossil fuel use, they are not necessarily optimal means of carbon reduction. Convenient public transit along rural commuting corridors could cut daily driving miles and reduce the number of two-car families.

It is worth noting, however, that as long as New England's electric grid relies on current levels of natural gas, a hybrid Prius may have a smaller carbon footprint than an all-electric Nissan Leaf, on account of the Prius's smaller battery and the Leaf's reliance on dirty electricity.[24] Like Priuses and wind turbines, even fully electric cars cannot be assessed for their value to the climate without considering energy used in their manufacture, maintenance, and place-specific application relative to alternatives. Again, the point here is not that electric cars are worse than gasoline cars; the point is that electric cars are part of a slow energy transition that could be accelerated dramatically by greater investments in mass transit.

Contrary to what we think we know about wind turbines, adding them to the existing energy grid at every marginally viable site does not produce a uniform set of ecological impacts. Just as a Prius achieves different efficiency on a congested city street than it does speeding uphill along the interstate, wind turbines are captive to the potential of the wind, which ranges from strong and steady to intermittent and volatile.[25] Northern Europe's offshore turbines synched with hydroelectric baseload power are an optimal arrangement, as they operate nearer maximum efficiency and offset more carbon thanks to strong and consistent wind speeds.[26] These offshore turbines displace no carbon-sequestering forest and require no idling fossil fuel power plant to take over when the wind ceases because hydropower can be dispatched from a thousand miles away. In less optimal scenarios, turbines' carbon reduction potential is reduced. One onshore wind turbine requires nine hundred tons of steel, 2,500 tons of concrete, and forty-five tons of plastic.[27] Each of these materials currently relies on carbon release. More specifically, when we consider turbines located in marginal wind corridors, requiring deforestation and road building, subject to periodic curtailment, and requiring fossil fuel power plants to idle on standby, the case for wind electricity weakens. These are the factors contributing to the ecological uncertainty now recognized by so many wind opponents across Northern New England's forested mountainous terrain.

We have been reluctant to hold wind turbines to any comparative scrutiny, but neither their form nor their function are uniform across places. Modern turbines fit into systems of electricity distribution, ownership, and governance differently. If we can recognize that Disney's Epcot Center is not just a giant golf ball, roller coasters are not miniature trains, and electric locomotives can be fueled by coal in a faraway power plant, then we should be able to grasp that wind turbines' familiar form belies great complexity.

The argument that wind turbines symbolize so much good that they divert attention from their shortcomings resonates closely with environmental sociologist Jesse Goldstein's study of venture capital directed toward clean technologies.[28] Even technologies intended to be disruptive can make only small dents in climate change by the time they are applied in the profit-driven ecosystem of industrial capitalism. Lightning rods for public hopes and dreams, Goldstein calls such technologies "non-disruptive disruptions." Their branding, he observes projects unabashed but undeserved optimism, focusing on discrete effects that "can be cast as unassailably good, green, and desirable."[29]

Renewables in an Unequal Society: Another Symbolism

Without question, the energy transition is also slowed by people who hate wind turbines, but again, that hate derives from things that turbines symbolize, not necessarily what they do. In a room full of Renewable Energy Certificate experts at the 2017 Renewable Energy Marketing conference, I listened as a few of them made small talk before a workshop began. They were bemoaning the resistance to renewables they encounter in conservative Western states. As they spoke, I imagined the remote parts of Wyoming I have driven through. I thought back to one particular desolate interstate gas station near a pit mine, with a trailer court where workers rented housing by the week. There was nothing around for miles, except the mine and that gas station. It was hard not to reflect on how culturally distinct the communities in those places are from where we were sitting in Manhattan's Upper East Side. As a rural sociologist and a scholar of inequality, I understand why renewable energy is unpopular in many communities. But hearing the two REC brokers express true bewilderment, it occurred to me that rarely does anyone acknowledge the particular moment in history in which we are trying to wean ourselves off fossil fuels or how that context impacts our progress.

Energy companies have made us all addicts of their cheap supply, but much of working America is far more susceptible to those companies' transition narrative because of inequality. There is of course the threat of renewables raising our energy bills, but there is also something about wind turbines spinning freely in the desert, making money for invisible people far away, that might not sit right with everyone—not just because of "Big Wind" per se, but because of broader economic changes that wind turbines and solar panels symbolize. We know that in communities dependent on fossil fuel extraction, market ideologies, local identities, and cultural animosity play heavily in engendering resistance to renewables.[30] But those traits may have just as much to do with fear of uncertainty than with allegiance to the company and its product, which would mean that these traits exist across much of rural America. It is reasonable, and overdue, to ask how the slow energy transition is aided by a legitimate suspicion of how ordinary workers, not just fossil fuel workers, are supposed to fit in to the future labor force.

Whereas the physical dynamics of fossil fuel extraction demanded scores of workers in high-risk, high-pay occupations, renewable electricity jobs are different. Petroleum extraction's reward structure privileges workers because they are essential to securing a containable, portable, and valuable commodity in global demand. The wind and solar sectors are carving out niches for profit around natural resources that cannot be contained, transported, or owned. After installation, renewable energy may require relatively few workers, laboring in conditions that are more favorable and thus earning less pay than in mining, drilling, or staffing power plants. In fact, the pressure is on renewables research and development to devise systems that minimize the need for human labor. The goal of R&D is to deliver technologies with the lowest possible long-term operating costs to maximize appeal to potential investors. At a time when most rural communities desire new employers offering middle-class wages to hundreds of workers, wind turbines offer something different. Their systems are monitored remotely, and maintenance is infrequent enough that laborers may not live nearby. In fact, a 2019 BBC article spotlights a team of scientists designing drone technology to monitor and eventually service wind turbines, which puts the very few jobs promised by the wind sector on the chopping block not long after they came into existence.[31]

Though highly touted by politicians, many green industries are themselves emblematic of industrial restructuring because of their lean operations driven by short investment horizons. While the media applauds explosive growth in the green jobs sector, no one digs into the details about which jobs count as "green," how they pay blue-collar entry-level workers in salary and benefits, where these jobs are, and how long they last. Realistically, a national green jobs industry will do little if it is not widely distributed across places and levels of skill. If renewable energy jobs are based in cities and suburbs or require constant travel, then the rural communities hoping for the renewable economy to throw them a postindustrial lifeline will have little to celebrate. Moreover, as I learned at the conference in New York City, the renewable energy economy is now integrated into the fabric of modern finance. Thus, the stratification of renewable energy's payroll looks similar to other postindustrial sectors, offering lucrative salaries to engineers and bankers but little hope of reviving a blue-collar middle class. Communities that have experienced restructuring before are already familiar with nominal "retraining" programs in clerical or technical fields that rarely replace the status and pay of their old jobs—if "retraining" offers any new qualifications at all.

In a world where rural restructuring has persisted for decades, wind turbines, with their networked and automated systems, are symbols of uncertainty for the average worker, just like the smart electricity meters that replaced the need for human meter readers and touched off a veritable revolt against American electric utilities. Some utilities deployed smart meter technology without any consideration of equity or transparency. Stories abound of utilities allowing consumers' monthly bills to double or triple after smart meter installation, with no publicity campaigns about the switch to demand-side pricing.[32] Moreover, there is no evidence that anyone considered that it is society's most vulnerable who are in their houses during the daytime, paying higher electricity rates to accommodate peak demand from businesses, many of which pay lower rates already. More broadly, customers of loosely regulated, historically corrupt energy systems have little reason to trust that smart meters, which only disclose energy prices, are not obscuring new tactics to manipulate those prices. These issues highlight that technology may do little for equity if we do not move more decisively beyond the passive customer model of energy commerce.

In summary, renewable electricity foretells an ominous outcome for working-class Americans. Wind turbines are not merely energy sources with new economics. They are energy sources for a new economy, an economy already defined by inequality that reaches new heights seemingly every year and in which automation and mechanization are eliminating the need for human labor with increasing speed. Already, widening inequality renders modern workers less able to afford housing, childcare, health care, home heating, education, and transportation, much less save for their retirement. By the numbers, renewable energy may benefit lease-holding landowners and wealthy business people more than ordinary Americans and their communities, which does not bode well for renewable energy's popularity and raises questions about fairness that transcend typical concerns about siting. If we cannot ensure that green jobs can compete with dirty jobs, we risk making green industries politically unpopular, further highlighting the need to meaningfully integrate environmentalism with social justice.

To understand renewables resistance, research should interrogate how modern levels of inequality make workers everywhere vulnerable to shifts in how society generates and uses energy. Frameworks that promote justice amid the energy transition should consider how energy justice is promised, achieved, and manifest in all contexts. This includes confronting the lived realities of a potentially hollow green labor force and what it represents for the broader working public, not simply those displaced from the generous payrolls of fossil fuel companies. Accounting for inequality is essential for contextualizing resistance to renewable energy, revealing that it is a rejection, not necessarily of climate change, environmentalism, or urban elites, but of broad labor force changes symbolized by structures with absentee owners and little need for local workers. This negative symbolism seems as powerful as that which celebrates turbines for their perceived benefits.

Conclusion

Renewable technologies and the energy systems they power are sociotechnical systems. Historically, the interests profiting within those systems have benefited from creating docile, passive, and entitled consumers. The creation of such consumers has allowed society's dominant transition discourse to be scripted by and in service to those drilling, extracting, and transporting natural gas; those manufacturing, installing, and investing in wind turbines; and those propping up the grid-scale electricity infrastructure.

Given winds' variability in strength and consistency, and wind projects' variations in backup power, regional carbon footprint, policy architecture, transmission capacity, construction and transport requirements, and other ecological trade-offs, it is easy to understand why some backyards may be better suited for the technology than others when accounting for greenhouse gases. To be clear, the problem is not that wind turbines are universally ineffective, but rather that they are a socio-technical system that is not uniformly suited for every geographic context to reduce greenhouse gas emissions.

With resonance as totemic icons in a widely publicized and widely accepted transition narrative, turbines serve as prominent symbols of ecological responsibility that affirm environmental progress and our citizenship in an environmentally progressive community. As we have seen in Northern New England, to oppose wind turbines is to lose one's environmental credibility. With widespread unquestioning support projected onto their physical form, Northern New England's ridgeline turbines evince a superficial zeal for planetary improvement that extends to all of us projecting our hopes and dreams for a cleaner future onto objects.

It is second nature for us to associate wind turbines with nature, grace, and purity, but staking our identities to them means we stretch our personal comfort in critiquing them. We are loath to concede that icons, despite our hoping and dreaming, put us on a trajectory of improvement so slow that one must wonder if American policy makers have ever read a newspaper, much less any climate research. Seen another way, renewable icons let us off the hook for not trying harder to change our systems, by constraining society's imagination for alternative paths to sustainable energy. Because the fossil fuel industry's influence over national and regional energy transitions suppresses viable alternative models and ignores the ticking climate clock, we must acknowledge ways in which "transition" narratives reveal a highly manipulated energy reality, similar to but more pervasive and consequential than typical greenwashing.

As progressive consumers who often shape narratives, scholars must question whether our own work treats turbines as inherently good, because the status quo energy reality envelops us, too. Have we considered that the way we think about wind turbines is shaped within broader structures of knowledge and influence that privilege some renewable energy solutions over others? Working toward energy systems more rooted in justice and sustainability requires making those things central to our conversations, instead of what is being prepackaged by hegemonic interests.

To ensure we use resources more sustainably will require a higher level of engagement with energy policy. To that end, we have more to worry about than how well turbines work. We must consider the social and ecological sustainability of entire systems, and compare our renewable icons to potential alternatives. We also have to meaningfully engage with how our shifting energy regimes relate to deep and fast-widening social inequities, to ensure that the renewable-powered economy really does benefit communities with more than vague promises for "green jobs."

If we can take one lesson from this perspective, it is the importance of distributing the renewable industry's benefits more widely. Clean energy would be a much easier sell politically if the pathway were lit with significant and long-lasting benefits for places hosting renewables. In the case of wind and solar installation jobs, assembling complicated things made overseas sounds a lot like shopping at Ikea; the sense of accomplishment is nice, but the payout could be better. Why not build a better renewable manufacturing economy domestically, creating jobs less dependent on weather, with higher wages and upward mobility? And why should we be intimidated by predictions that a full electricity transition is too immense without relying on gas? A massive national buildout of renewable infrastructure and microgrids, tailored to regional climate goals, should only scare the gas industry. Certainly, the costs of such a transition should not scare voters if it entails employing millions of Americans in high paying jobs. Even better, community ownership of energy infrastructure could help rural families lower their bills, pay property taxes, and put kids through college.

8

Ecological Modernizations
or Capitalist Treadmills?
• •

It was a crisp fall evening when I attended the public meeting announcing Vermont's new energy plan, but the air in the school cafeteria felt stuffy and stale. (This is the meeting I described in chapter 2, where progressive suburban voters advocated for a carbon tax.) The stuffiness made me wonder if the school had switched on its furnace early, but near the end of the evening's proceedings—as the presenter described the efficiencies of cogenerating electricity and heat—I noticed the fluorescent light bulbs blazing overhead. I wondered why a public building in the most affluent corner of the most environmentally progressive state in the United States had such inefficient lighting. This thought led me through a string of internal questions about the feasibility of executing Vermont's ambitious energy goals given its confounding environmental track record.

I learned that night that Vermont's 2016 energy plan is the state's first effort to link energy consumption to land use, which makes the plan long overdue. Much of the country's experience with rural land use planning is fraught because early efforts were meant to maintain rural character by enforcing minimum lot sizes. That approach worked, but it also drove rural sprawl and accelerated dwindling affordability. Thanks to rural zoning, Northern New England's rural character offers such natural isolation for homes that one might never guess the number of people living in a munici-pality by driving on its paved roads. But local school principals know too well. They hand over much of their school budgets to oil companies in order to bus their students along circuitous road networks.

Throughout rural Northern New England, sprawl has proceeded slowly and steadily, making the region more dependent on cheap gasoline than its progressivism lets on. As in other states, there are large confinement dair-ies, and strip malls have filled in pastures and meadows to the detriment of downtowns and villages. Just a few years earlier, I had interviewed for a state job in an office park off the interstate, nowhere near public transit or even a bike lane. I love Vermont just as much as all the young people who have its outline tattooed on their bodies, but the state has a history of embarrass-ing contradictions.

Sprawl matters in the region's renewable energy debates. With less than 20 percent of the region's carbon emissions originating from electricity gen-eration, focusing so much on electricity instead of transit strikes many as disingenuous. While current trends suggest that electricity will be needed for heat and transit in the future, addressing climate change with the urgency demanded by scientists requires overhauling heating and transit systems, rather than simply converting them to electricity. For a region whose resi-dents have a reputation for thrift, frugality, and resourcefulness, Northern New England is leaving a lot on the table in terms of carbon. I contemplated this as I sat beneath the flickering, hot lights.

I am not the only one to grapple with Vermont's environmental duplic-ity. Journalists scrutinized former Governor Peter Shumlin, perhaps the loudest proponent of the state's wind development, for extreme personal contradictions. Upon taking office, Shumlin leased two gas-guzzling SUVs, Ford Expedition EXPs, for his daily transportation and was the first gover-nor in decades to enlist a small airplane to travel throughout the country's fourth-smallest state. In a state where hybrid cars abound, the governor's per-sonal transportation decisions disgusted many environmentalists. Imagine that you drive a hybrid car, power your home entirely from the sun, and are being schooled on renewable energy by a governor who apparently shows no personal concern for climate change. This is the reality for many wind opponents.

It can be hard to make sense of sustainability efforts accompanied by so many contradictions. How we reconcile the strengths and weaknesses of modern wind energy is a matter of perspective. Just as energy politics can be shady but normal and legal, wind energy can be good and bad, depending on how we look at it. This chapter draws from environmental sociology to reconcile the strengths and weaknesses of wind turbines and the general policy realm they come from. The chapter uses two perspectives encapsulated by different threads of research; one body of research is predisposed for optimism and the other is more critical.

Placing Ridgeline Wind in Theories of Eco-progress

A central research theme to grow from environmental sociology engages with environmental policies' material impacts. In the tradition of sociology's critique of capitalism, many environmental sociologists examine how profit motives co-opt ecological motives and overpower environmental policy's good intentions. Put simply, research asks how optimistic we should be that our policies can achieve their goals, given the pervasive influence of capitalism. Does ridgeline wind reflect a policy shift toward solving environmental problems, or are wind turbines artifacts of capital accumulation that worsen those problems?

The field of environmental sociology has given rise to two oppositional theoretical frames that offer useful guideposts for answering these questions: Ecological modernization theory (EMT), and treadmill of production (ToP) theory. One can think of EMT proponents as the optimistic camp, hailing from Western Europe, and ToP proponents as the more critical camp, working in North America. These theoretical perspectives are not mutually exclusive, and I will show that evidence from Northern New England supports both.

Ecological Modernization Theory

Dutch sociologist Arthur Mol's 1995 analysis of the chemical industry in Europe is credited as the first outline of the EMT perspective.[1] Mol's work observed a shift in how businesses treat the environment. At the time, most research in the new field of environmental sociology had emphasized ecological exploitation and crises.[2] Mol observed, however, a convergence of economy and ecology within management and governance in both the state and private enterprise. Thoughtful governance was bringing the ecological impacts of development into state decisions about the economy. EMT holds that this pattern of incremental and substantive changes amounts to a cohesive policy regime.[3]

Mol and his colleague Gert Spaargarten touched off a lively and decade-long academic debate, giving way to multiple attempts to articulate and rearticulate how the changes first observed by Mol should be understood globally. The crux of the debate has always been the limited evidence for EMT outside a handful of primarily Western European countries. American environmental sociologist Fred Buttel argued that EMT encompasses an "objectivist" strand, concerned with how businesses and states work toward integrating specific ecological principles and objectives into business and state functions, and a "social constructivist" strand that examines ecological modernization as a discursive phenomenon, carried through public narrative and requiring pressure from the ground up to actually compel change.[4] The first is "walking the walk," so to speak, and the second is "talking the talk." Australian political scientist Peter Christoff described ecological modernization as a phenomenon with either weak or strong manifestations, with strength deriving from widespread adoption and rooting in popular belief. Strong ecological modernization, Christoff argued, is ecological; institutional, systemic, and broad; communicative; deliberative, open, and democratic; international; and diversifying. Weak ecological modernization on the other hand, is economistic, technological, instrumental, technocratic, national, and unitary (hegemonic).[5]

While environmental sociology has largely moved on from debating EMT, I still find it useful for thinking through policy progress, or lack thereof, in the realm of renewable energy.

Northern New England's efforts to promote efficiency and reduce its reliance on fossil fuels reflect momentous change. Collectively, the region's portfolio of environmental initiatives offers evidence that state policy and institutional behavior can and do evolve in closer alignment with ecological principals. However, as I will describe, that alignment is inconsistent and periodic. The precise type of EMT on display in the region's energy policies is best classified as social constructivist and weak, highlighting clear room for improvement.

Addressing Global Climate Change. All three Northern New England states are doing more for the climate than just embracing wind and solar. All three states are signatories to the Regional Greenhouse Gas Initiative

(RGGI), the nation's first market-based regulatory program to reduce green-house gas emissions. The RGGI includes Canadian provinces and is one of only two such compacts in North America. Membership in the initiative gives the three rural states access to revenue from auctioning unneeded CO_2 emissions to the more populous states in the compact. With these and other funds, all three Northern New England states have subsidized electricity conservation efforts through programs that promote efficiency.

Although I have pointed out flaws in the state's approach, Vermont actu-ally leads the region, and perhaps the nation, in a legislative commitment to reducing electricity demand. The Vermont legislature established a Clean Energy Development Fund that invests in a wide array of programs and proj-ects that advance "cost-effective and environmentally sustainable electric power and thermal resources—primarily renewable energy and combined heat and power (CHP) technologies."[6]

Vermont has diverted electric ratepayer funds toward conservation efforts for two decades. In 1999 the state legislature authorized the Vermont Pub-lic Service Board (now the PUC) to charge electric customers, both com-mercial and residential, a fee according to their usage. The state also created Efficiency Vermont, an "efficiency utility" that administers an extensive portfolio of programs to assist ratepayers in the process of reducing their use of electricity. Efficiency Vermont is administered by the Vermont Energy Investment Corporation (VEIC), an independent NGO. Efficiency Ver-mont offers technical assistance and incentives for energy-efficient build-ing design, construction, renovation, equipment, lighting, and appliances. Efficiency Vermont's primary priorities are reducing the size of future power purchases, reducing the generation of greenhouse gases, limiting the need to upgrade the state's transmission and distribution infrastructure, and min-imizing the costs of electricity. In just seven years, the work of Efficiency Vermont helped the state became the first in the nation to realize a reduc-tion in electricity sales, declaring itself "load growth negative," a novel con-cept in 2007 that other states have sought to emulate.

Maine and New Hampshire accompany Vermont at the fore of energy efficiency legislation, though deregulation in those states softens their leg-islatures' authority to institute the same sweeping changes. Maine's effi-ciency efforts began as utility-administered programs in 1997, but entered the fold of the legislatively created Efficiency Maine Trust in 2010. Even libertarian New Hampshire followed in 2014, removing oversight of its decade-old efficiency programs from utilities and instead creating the

Energy Efficiency Resource Standard to maximize the potential for efficiency efforts to reduce overall electricity usage in the state.

All three states were also early adopters of policies to promote residential solar electric conversion. Along with tax incentives on the purchase of solar equipment, each state passed net-metering legislation around the year 2000 to compel electric utilities to purchase excess power from residential solar panels. Additional policies entice homeowners to invest in new solar installations by shortening the payback time relative to standard electricity rates. For example, all three states instituted property assessed clean energy (PACE) programs, allowing property owners to obtain loans for clean energy improvements that convey to new owners should the property be sold before the loan is paid off. This incentivizes efficiency improvements even for homeowners unsure how long they will stay in their home. The Northern New England states have also endeavored to spread the benefits of pro-solar legislation to ratepayers who do not have the privilege of homeownership. To summarize, Northern New England states have earned their progressive reputations through early commitments to energy efficiency.

Vermont: The Environmental State? Debates about ecological modernization engage the concept of the "environmental state"—that is, the evolution of a particular type of modern government that internalizes ecological goals with policies that advance sustainability. In this section, I will explore Vermont's progressive environmental accomplishments though the conceptual framework of the environmental state.

Some of Vermont's energy policies have impacted the way markets function, a key element of ecological modernization. For example, thanks to policies set by regional grid operator ISO New England, Vermont's achievement in reducing its electricity use allows the state to access additional revenue from participating in the forward capacity market. In essence, as power plants undercut one another to offer the grid the best deal on electricity, Vermont's efficiency utility can be compensated for its contribution to the reduction of demand, because the state is saving the grid money.

Furthermore, Vermont's Act 56—legislation enacted in 2015 that instituted a formal renewable portfolio standard (RPS)—signaled an aggressive shift away from fossil fuels by advancing the law's scope beyond traditional electricity use. Act 56 created a series of tiered mandates for electric utilities to sell renewable electricity to their customers. Apart from the RPS (tier 1), the mandates incentivize Vermont's electric utilities to promote "distributed" small-scale renewable projects such as microhydro and grid-tied residential solar projects from which the utilities can purchase excess supply (tier 2). Through the third tier, the legislation sets the stage for wholesale transformation of the state's carbon footprint by requiring electric utilities to reduce their consumers' fossil fuel dependence in domains outside residential electricity. For example, to meet tier three mandates, utilities can facilitate ratepayers' transition to electric vehicles or electric home heating, which are the state's largest sources of carbon emissions.

Policies such as this are feasible in a regulated electricity market where consumers are captive ratepayers to monopoly utilities closely watched by state overseers, but, as described in chapter 5, Vermont's policies are born from close collaboration with the state's largest monopoly. Thus, understanding Vermont's success in enacting such progressive policies requires understanding the utilities that policies regulate. The following paragraphs describe how Vermont's Green Mountain Power emerged as a national leader in renewables.

Green Mountain Power—Utility of the Future? GMP is state legislators' private-sector ally in their efforts to promote energy efficiency and renewable electricity. Serving the majority of the state's households, GMP is at the forefront of climate responsibility, grabbing national headlines for promoting an ecologically minded business model that would make most energy executives queasy.

GMP's headquarters sit in an industrial park behind several large warehouses at the end of a long and winding cul-de-sac outside Burlington. The building is unremarkable from the exterior, save for two large solar arrays and several electric vehicle charging stations. The lobby hosts a collection of historical electric meters and transformers of various sizes with placards explaining their function and uses. One transformer is painted yellow and has been converted to a receptacle into which customers can deposit their paid bills. Three of the lobby's walls are glass, and when looking straight ahead past the glare from the reflection of the world outside, one can see the enormous illuminated control center of GMP's statewide power operations. This fishbowl view lends a sense of transparency, which one would hope to find in their state-protected energy monopoly.

In addition to wind, GMP manages a renewables portfolio that includes biodigesters generating electricity from cow manure, to which ratepayers can opt in, as well as the state's largest solar facility and medium-sized hydroelectric dams. GMP's solar portfolio is also noteworthy for its inclusion of community solar. After a partnership with a "sharing-economy" solar start-up failed in 2015, GMP partnered with a local company to install solar panels on the roof of one of its buildings explicitly for local low-income families to buy into so that they could reap the tax benefits of owning and using solar.

What makes GMP most exceptional is its general orientation toward innovative trends in electricity generation and distribution—trends that most electric utilities find disruptive, if not fundamentally threatening to their business models. GMP is positioning itself to facilitate efficiency well into the global transition away from fossil fuels—not simply by ramping up its portfolio of renewables, but also by rethinking the design of regional electrical systems. According to public interviews with former CEO Mary Powell, GMP envisions a future in which the interconnected energy grid has given way to a fractured network of stand-alone and community-scale power systems, relying on connectivity to a large grid only for ancillary electricity. GMP is therefore positioning itself to help this transition along by building the technical expertise and logistical capacity to build and service this futuristic type of power system.[7] GMP was among the first electric utilities in the country to seize upon Tesla Motors' power wall technology, and GMP has since allowed customers to use other manufacturers' batteries as well. With remote access to control smart battery systems installed at its customers' homes, GMP has a carbon-free source to draw from under peak load conditions.

When I spoke with GMP's director of community affairs, they made it clear that the company's outlook on the future of electric utilities is not representative of GMP's industry peers:

Since we announced our partnership with Tesla, I've talked to more utilities around the country, curious, like, how are you doing it? Why? What are you doing? Some for the right reasons, because they want to just do better by their customers and for their customers, some, because they want to defend against it, which is totally the wrong approach in my mind. I would say, yeah, we're definitely not the norm when it comes to how we're thinking about the future. We're looking at it in examples of, hey, what happens if half of our customer base essentially disappears because of rooftop solar? What does it look like, and what are the other things we need to do to manage that? Yeah, that's the thing about Vermont. We're tiny. . . . We're a small utility, we're a small state, but doing things like this can have a big impact as a percentage of the state, and it's a great place to showcase this stuff, I think. Then we have a regulatory body and a governor and an administration that wants to see energy innovation happen here, so that's huge.

My informant in the legislature returns this admiration, describing GMP as a strong ally and, in some cases, the motivating force behind several of Vermont's progressive policy achievements. My informant reflected on time spent as a member of the Natural Resources Committee:

> In fact, a lot of the ideas that I got and that lead to, for example, the solar feed-in tariff . . . were because Green Mountain Power recognized the benefit of having lots of distributed solar generation which are putting out the most power when they are facing the possible need to buy expensive market power, so they could avoid that peak power hit. They were encouraging distributed generation, paying more for it, solar. We got that idea from them and made it apply statewide. We were lucky in Vermont in terms of how easy is it to pass these things. We didn't have any utility opposition except from small communities. Green Mountain Power is a progressive utility. . . . They recognized that no matter what they do the world is going to change. You'll see utilities in the rest of the country fighting it, metering as far as they can. Green Mountain Power's attitude is we need to be green and reliable and affordable. Green is an equal leg on their stool. Like Germany, they see that there is opportunity in embracing this new paradigm and taking control of it.

So why is Vermont's utility landscape so progressive? In part, the reason is that the entire region is charting a progressive electricity course. Recent renewable initiatives initiated by the legislatures of New York, Massachusetts, and Rhode Island are a testament to a broad Northeast political progressivism, which the Northern New England states have fostered with particular intensity. Even so, Vermont's main electric utility seems exceptional. Rather than simply buying low cost hydro-electricity from Quebec, GMP claims that it plans to upend its entire business model. Arguably, the company's progressivism mirrors that of its customers.

If the case for ecological modernization could be made in the United States, it would most likely be built around evidence from a handful of states—probably Vermont; California, the nation's renewables and pollution-control pioneer; and perhaps Oregon, America's most aggressive regulator of urban sprawl. Scholars debating whether environmental milestones constitute evidence of ecological modernization look for depth and breadth in changes to markets and institutions. Put simply, scholars distinguish transformative social and economic change from mere superficial modifications meant to symbolize change. We ask: Are state policies, market governance, and the public conscience being refashioned to account for ecological harm, even in instances when doing so poses a threat to the capitalist status quo? In what follows, I first elaborate on the reasons for EMT optimism and then detail reasons for a more critical perspective.

The Case for EMT Optimism in Northern New England. Compared with other states, Vermont, Maine, and New Hampshire have made three significant strides in changing the electricity business. First, state policies are starting to change the way electricity markets function by incorporating ecological considerations and negative environmental externalities. All three states mandate that electricity suppliers participate in a regional cap-and-trade market for carbon emissions as signatories to RGGI. This mandate is a textbook example of bringing environmental externalities into the economic balance sheet, from which they are typically excluded. Further, by bringing efficiency gains to the forward capacity markets, the state of Vermont has secured a revenue stream to continue making progress in efficiency, which will in turn secure additional revenue so long as growing use of electric cars and heating does not erode the strategy.

Second, the policies enacted in Northern New England represent significant penetration of the environmental sphere into the apparatus of state government. All three states now reinvest proceeds from their emissions auctions into energy efficiency programs that are run independently of the profit-seeking electric utilities. These "efficiency utilities" serve as clearinghouses for numerous energy conservation programs available to private citizens, businesses, and NGOs. In addition to the re-engineered carbon and electricity markets, these new structures and institutional arrangements represent explicit changes to the legal and economic architecture that once treated the environment as a limitless and free resource.

Third, the very nature of public debate triggered by renewable installations offers evidence of a reflexive dialogue, in which energy's consequences and society's potential sacrifices are discussed openly for the first time since nuclear power entered the region in the 1970s. The public outcry over every wind project creates ripples that spread to other places and compels a growing cross section of the region to learn about electricity. Big wind and solar projects united diverse local interests to think about energy as both a local development issue and a divisive social justice issue with far-reaching consequences. The activism that emerged around ridgeline wind helped drive new energy legislation and led to tighter siting restrictions to protect the neighbors of future wind installations. Newly proposed wind developments face a better-informed public, helping to change the calculus that developers and utilities use to weigh the pros and cons of proposed projects.

Put simply, it is easy to find irrefutable signs that Northern New England's environmental policies resonate with EMT. However, each of these signs is counteracted by concerns about ridgeline wind turbines that poke holes in the case. Those concerns call into question the efficacy of renewable energy markets, the penetration of ecological thinking in state apparatus, and the depth and resonance of reflexive dialogue to have emerged from energy controversy.

New England Is Not New Europe: How EMT Falls Short. Examining New England's wind industry in tandem with broader changes to electricity markets paints a fuller picture of the motivation behind major structural changes in utilities and the way they interact with government. In a nutshell, Renewable Energy Certificate markets make utilities, not more ecologically focused, but instead more capitalist than ever—and it seems unlikely that this is a kinder, gentler, greener capitalism.

A common criticism leveled by wind opponents is the industry's growing scale and the mechanisms through which it profits. Contrary to the tenets of EMT, ridgeline wind turbines do not represent changes in the relations of energy's production. Rather, the turbines are a doubling down on traditional large electricity systems and market structures. This doubling down is evident in the corporate electricity consolidation to sweep all three states. Consolidation, as in any business, helps ownership realize efficiencies of scale. In the case of Northern New England's electric utilities, consolidation has occurred through takeovers and mergers, resulting in ownership of the region's largest electricity providers moving out of the region and, in the case of Maine and Vermont, outside the country. This process results in job losses and in profits and spending being siphoned away, and remember that grid maintenance is slipping, too. These changes paint a portrait of the electricity business as usual, not structural changes carving out greater space for ecological concerns.

Ridgeline wind does little to overhaul the institutional structures that make electricity markets ecologically damaging in comparison to residential solar feed-ins. Turbine installations and now many large solar developments represent an energy market in which rural landscapes host the investment capital and electricity capacity of faraway cities, with some neighbors receiving little benefit or say in the matter. The first feed-in tariffs for residential solar owners, by contrast, flipped production and consumption roles entirely. Consumers were allowed to become producers and were granted long-term price protections.

Overall, Northern New England's political economy of energy allows utilities to be bottlenecks to greenhouse gas reductions. David Hess writes that ecological modernization is limited by utilities' orientation toward growing profits: "In the large corporate and financial organizations, the technological and discursive shifts associated with ecological modernization take place within a context of investor interest in continued profitability. Change in their paramount value of profitability growth would require significant shifts in state regulations so large, publicly traded corporations would be required to set general societal benefit concerns above the interest in maximizing profits for shareholders."[8]

Even nationally celebrated Green Mountain Power fits this mold. After securing electricity from two large wind installations developed with multinational capital, in 2016 GMP turned down the option to purchase electricity from smaller wind projects under development by Vermont residents. In local media, counsel for one project attributed GMP's decision to the fact that it is a subsidiary of a foreign natural gas company also expanding gas lines within the state. GMP's CEO balked at this accusation and attributed the company's position to the proposed project's cost to ratepayers, a concern shared by the state's smaller electricity companies and cooperatives. Regardless of GMP's motivations, however, the region's experience with wind energy has been attenuated by corporate needs rather than local will.

While Northern New England's environmental policy achievements demonstrate that progressive eco-politics exist on both sides of the Atlantic, the region's experience with wind turbines should be viewed differently from the European experience. Northern New England's experience contradicts the argument of British political scientist David Toke that our green identities are greening our energy institutions.[9] Identity may be central to shaping feelings about wind turbines in the United States and vice versa, but citizens who do not work in the wind industry play a passive and limited role. Toke and many others cite Danish farmers who pioneered the modern wind turbines on rural island communities facing exorbitant electricity rates as evidence of such personal engagement. The Fox Island wind project in Maine's Penobscot Bay offers a parallel story. It was a hard-fought, grassroots, community-driven project to make electricity more affordable for the islands of Vinalhaven and North Haven. But wind projects on inland ridgelines grew from different scenarios that offered fewer and weaker opportunities for local engagement. Hardly driven by local activism, some ridgeline projects take neighbors by surprise.

Northern New England wind businesses are structured differently as well. Toke describes Europe's legacy power corporations, renewable energy companies, environmental NGOs, and regulators as discrete and independent actors operating transparently. In Northern New England, as in most of the United States, we see that policy architects—including utilities, developers, distributors, politicians, and NGOs—all overlap, which motivates them to align and fracture in ways that are decidedly not transparent. Recall that Vermont has a "community" wind project owned in part by a foreign gas company, which is owned in part by a notorious fossil fuel conglomerate. The community has zero ownership stake in the "community" project.

Focusing attention on the ways in which ridgeline wind turbines fit within larger systems of capital accumulation helps temper the argument that wind energy is something modern and progressive with the argument that wind energy is actually something old-fashioned, just newly packaged. This reflects what Hess has called "an intense game of cooptation" by which dominant industry players use renewable technologies strategically to increase their profits, without prioritizing environmental improvements.[10] In what follows I describe how Northern New England's ridgeline wind resembles this co-optation using a more critical theoretical frame that emphasizes the magnetic pull of growth in shaping society.

Windmills and Treadmills

For my informant Karen, whom I met outdoors at a coffee shop as autumn beckoned in late August, it is easy to see wind turbines as engines for a treadmill. Karen sees turbines as parallel to other development in the region, which related or not, both signify an increasingly consumerist society:

> People are continuing to develop and consume because the electricity is available. People are continuing to develop and create electricity because the subsidies are there and they can make a hell of a lot of money off it. Those are

both economically driven motives and they complement each other. . . .
I think, on a deeper level, they're both interrelated by the fact that they are
engendered by the same culture, which is one of consumption and disregard
for the natural landscape and . . . entitlement that we can do this to the
mountains because we need the electricity and we need to maintain our
standard of living.

Karen is in her forties and works in education. She has lived in the region
for around twenty years, but she had been a regular seasonal resident during childhood. She hits on a point about consumers that Ralph made too,
earlier that summer. He told me, "The standard approach these days in
renewable energy, in any of these arenas, is to build more generating capacity, and then to claim carbon benefits. And that's total bullshit. We need to
start rebuilding changes up here." When he said "up here," Ralph was
pointing at his head. He was advocating a rethinking of electricity and
consumerism, to sustain the natural systems on which humans rely rather
than blowing them up. In what follows, I carry Karen and Ralph's perspective out further, showing how industrial wind turbines and the policies
from which they have emerged fuel expanding consumer demand for
electricity.

Environmental sociologists agree widely that capitalism has a poor environmental track record. There is less agreement around questions relating
to how easily ecological considerations can be introduced to markets. Allan
Schnaiberg's explanation for why capitalism so consistently supersedes concern for the environment uses a treadmill analogy.[11] Capitalism is predicated on growth because capitalists are pressured to increase production so
that lower prices, driven down by competition, do not result in lower profits. The expectation of growth extends beyond manufacturing to other
actors with financial interests in production. Bankers and shareholders also
benefit from growth and are complicit in promoting it when they lend, bor-
row, and invest. Municipalities push for population and infrastructure
growth as a means of securing political power, tax revenue, and prestige. In
early 2020, the world watched capitalism's reliance on growth play out in
real time on stock markets as COVID-19 rattled investors' optimism that
markets will continue to expand in the near term. As the pandemic plays
out, stock prices have rebounded but still hinge on signals about the likelihood of growth, like vaccine rollout and the prospects for a quick economic
recovery.

Diverse growth interests collectively exert forward momentum under-written by ecological systems. The food we eat, the fuel we burn, and the materials we buy and throw away are physical things that form the back-bone of our growth economy, but little or nothing is done to ensure we source them sustainably. It is because of this steadily operating treadmill that meaningful environmental progress has remained elusive. Acknowledging ecological harm slows the treadmill, and slowing the treadmill will ulti-mately cost somebody, somewhere, financially.

A related theory incorporates consumers to the treadmill analogy. While producers are on a treadmill to accumulate capital, consumers engage in a perpetual cycle of accumulating belongings, experiences, and statuses.[12] Like investors, consumers' expectation that the future will be brighter and more bountiful is predicated upon the expectation of growth. More people with more money mean more customers making more purchases, using more natural resources to create more products, and eventually amassing more wealth to save, spend, or reinvest. Together the treadmills of production and consumption fuel the economy and exert tremendous demand for real fuel, which comes at increasingly high environmental costs. This was clearly evi-dent during shutdowns from COVID-19. When people were not leaving their homes and buying things they did not need, and when factories sat idle and highways were empty, air pollution plummeted. Estimates of the benefit to the environment from the pandemic's negative growth are staggering.

The Electricity Treadmill

Wind turbines are a treadmill of their own. As socio-technical systems, wind turbines are brought into being by a vast corporate matrix hoping to make more and more of them. Chapter 5 examines how they function as invest-ment vehicles for large institutions and shareholders, but the industry that designs, builds, and sells wind turbines is its own industrial juggernaut. A few competing firms produce grid-scale wind turbines, and that competi-tion creates the incentive to produce them bigger, more cheaply, more effi-ciently, and in greater volumes and with better monitoring, service plans, and warranties. When critics of wind projects reference "Big Wind," they are talking about the global web of corporations, including component sup-pliers, lobbyists, consultants, contractors, and construction firms, that are linked in helping producers of grid-scale turbines promote their products and thus grow their profits. Production is in no way governed by analysis of where wind turbines will be most effective or how many should be installed in these places. It would appear that the only assumption underlying production is that large turbines can be placed anywhere that the wind has been observed to be profitable, and that making them even bigger means that more places can be profitable hosts. While I do not intend to question the ecological merits of renewable electricity relative to fossil fuels writ large, it warrants observing that a large and growing industrial wind sector functions no differently from capitalist enterprise in any other sector.

In its totality, the wind turbine treadmill does little to slow society's appetite for electricity because electric power is its own treadmill. Newer, cheaper forms of electricity spur growth in society's demand for electricity, and, of course, the companies that sell electricity increase their profit by selling more. To the first point, society is confronting the paradox of its efficiency: the more we save, the more we have to spend. That is, improvements to efficiency can be negated by growth in the number of appliances we purchase and power on. As efficient lights become cheaper, we install more lights. As digital displays become cheaper and more efficient, we install more screens—in the form of tablets, picture frames, restaurant menus, in-car televisions, billboards of every size, and now refrigerators with LED displays. Most office workers spend the day on their computers. Schools use smart boards and projectors instead of chalk and markers. As our summers become hotter, we install more air-conditioning. More air-conditioning, in turn, makes our urban neighborhoods even hotter by exacerbating the heat island effect, demanding more air-conditioning still. Every e-mail and digital photo we save is sitting on a server somewhere using energy to stay accessible and not overheat. Thus, despite greater efficiency in residential energy, the United States has seen a net increase in its overall energy demand, and Europe has seen half of its efficiency improvements negated.[13] Though Vermont deserves praise among its peers for using efficiency to reduce its electricity demand, the state has been aided by its stagnant population growth. Like that of all cold rural states, Vermont's electricity demand is poised to soar as households replace oil furnaces with electric heat pumps and gas automobiles with electric automobiles.

This persistent march toward excess amid an ongoing climate crisis demands that we think about the architecture of the systems on which we have come to rely. We can attribute our reliance on one centralized electricity grid to path-dependence, but there are important dimensions of corporate power and consumer lock-in to consider. Environmental sociologist Elizabeth Shove uses air-conditioning to illustrate how industrial sectors are constrained in their environmental innovation by the paradigms governing their production model and their products' use. Shove argues that improving the efficiency of individual air-conditioning units may be orders of magnitude less effective than rethinking how we cool indoor spaces—for example, cooling more spaces with fewer machines, or swapping in more efficient technology altogether, such as passive design, geothermal, or evaporative cooling—but we will continue with the status quo because of the considerable investments made by private enterprise.[14]

Shove's sobering observation seems particularly salient for understanding why climate policy so consistently targets electricity instead of more polluting industrial systems. State efficiency programs that divert funds to consumer products and services are still promoting consumption, whether refrigerators or subsidized twenty-dollar lightbulbs from China. The most effective means of reducing carbon emissions, like expanding mass transit and neighborhood heating and cooling, face very high political and practical hurdles. Other important sources of greenhouse gas, such as methane emissions from natural gas fields and animal agriculture, fail to even register on policy makers' radar. Redesigning these paradigms of production and use would disrupt major capitalist structures, including those related to petroleum drilling and distribution, and those sustaining the cultivation of corn and soy, which for decades have been interrelated industries. By contrast, adding wind turbines to the existing electricity grid is not disruptive of the treadmills in the least. It is adding a treadmill to a treadmill. It is akin to applying efficiency standards to air-conditioning units instead of cooling our spaces more intelligently; it is popular with the business world but less than optimal for addressing climate change.

Reconciling EMT and the Treadmill Perspectives

Relative to what has been achieved in other U.S. states, the successes of Maine, New Hampshire, and Vermont in implementing climate-driven policies and programs makes the states national leaders, but the extent to which their policies actually challenge the forces that drive environmental problems is questioned by many wind turbine opponents. Ecological modernization theory and treadmill of production theory offer complementary theoretical perspectives for making sense of Northern New England's experience with grid-scale wind turbines.

So should we harangue wind turbines as instruments of the polluters or celebrate them as evidence of ecological modernization? The answer is a matter of perspective and scale. Evidence of ecological modernization need not refute evidence of the treadmill, but theorists have argued that ecological modernization, in effect, reshapes the treadmill's consequences. That is to say, ecological modernization theorists maintain that capitalism can be regulated or redesigned so that economies sync with ecological systems and principles. A defining characteristic of ecological modernization is the emergence of reflexive institutions that mitigate their environmental impacts through conscious reforms and the development of the environmental state, which incorporates ecological principles into policy.

Many of the region's policies show movement toward more reflexive governance, indicating that indeed Northern New England, and Vermont in particular, exemplify ecological modernization in some domains. For example, strong consensus about environmental protections and climate change suggest that the region's policy makers demonstrate far greater reflexivity than those in other parts of the United States. The best example is Vermont's 2015 energy legislation mandating that electric utilities manage their ratepayers' eventual transition to renewable energies in transportation and home heating, the state's most carbon-intensive activities. As the state of Vermont's primary utility partner, Green Mountain Power is itself engineering a radical business model that promotes distributed generation, microgrids, and residential solar storage.

For some, the sticking point of EMT is the limited degree to which one can observe a pattern of ecological reflexivity spreading and advancing over time and across institutions—a modernizing *process*.[15] It is easy to recognize monumental changes in institutional practices that will reduce human impact on the environment, but it remains difficult to attribute a plurality of those changes to a cohesive strand of environmental logic or a growing ecological awakening. The move to rid our food supply of carcinogens and antibiotics, for example, derives in large part from consumer health demands. This shift has not altered the ecologically devastating scale of agricultural production processes. Northern New England's wind turbines, while appearing ecologically pure, have a complex origin story that never included a mandate to actually be pure.

With regard to Christoff's categorization of strong and weak ecological modernization, it would seem that Vermont falls into the "weak" category.[16] The state's wind turbines epitomize an economized technocratic approach to renewable energy. Vermont's deregulated electricity market allows it to make top-down changes with minimal buy-in from residents, and the state's policies rarely conflict with growth-centered economic priorities. And yet the state's clean, green image is legendary, which speaks to Buttel's distinction between objective and socially constructed ecological modernization.[17] While states like Vermont have succeeded at sustaining a discourse with strong narratives and powerful symbols of ecological reflexivity, the extent to which ecological reflexivity has reoriented the state's approach to environmental policy remains unclear. Vermont's policy path unfolds as discrete, incremental progress in niche environmental sectors, rather than broad reorientations toward ecological principles. In the following paragraphs, I describe an unfolding environmental disaster, which suggests that the battle between the economy and the environment, and thus renewable energy policy, is the same in Vermont as in any other U.S. state, that balancing growth with ecological health makes slowing the treadmill very hard amid rampant inequality.

"Teflon" Town and the Politics of Reflexivity

In 2016, news broke of contamination in Vermont's southwest corner around ChemFab, a plant that once coated fabric with Teflon—the material known for nonstick and stain-resisting properties. Several hundred residents learned that their wells were contaminated by four decades of particulate perfluorooctanoic acid (PFOA) settling in the wind shadow of ChemFab's three smokestacks and leaching into groundwater. Though ChemFab closed in 2002, several residents showed high concentrations of the substance in their blood fourteen years later, and after a year of monitoring, PFOA concentrations in some wells continued to rise. This case has drawn intense scrutiny to state regulators and brought to light several reasons to be wary of Vermont's case for ecological modernization. Foremost among them is evidence that suggests that even where strong environmental convictions pervade the electorate, the state's capacity to protect its citizens is stymied by a political culture that shows consistent deference to business.

Exemplary reporting by the *VTDigger*, in a series titled "Teflon Town," traced the ChemFab calamity back to the 1960s, when the company's president claimed that its materials were harmless.[18] ChemFab workers recall forgoing gloves and masks when handling toxic substances. They were instructed to save incineration for the night shift, when soot would not be seen or smelled by the plant's neighbors. But the most jarring revelation is evidence of lax oversight from Vermont's environmental regulators, who pandered to ChemFab executives in fear they would move their facility out of state. Reporters detail how hundreds of air quality complaints by the public went ignored, along with recommendations that regulators learn about chemical toxicity and follow recommendations made by regulators in neighboring New York. Consistent environmental violations resulted in minimal consequences.

Given the opportunity to reflect on their job performance, former state officials, while contrite, invoked Teflon qualities themselves. Evading any culpability, they rationalized a political culture in which jobs were paramount and environmental regulations an afterthought. One former state official is quoted as saying that "tension" between government's environmental and economic arms is "healthy." I suspect that sickened neighbors find no humor in that choice of words. Scientific analyses and public health assessments failed to take precedence over ChemFab's status as an employer paying decent wages. For officials interviewed by the *VT Digger*, that these workers earned high wages, paid taxes, and buoyed the fragile economy explains why they are now drinking the same chemicals they once applied to fabrics.

The "Teflon Town" series touched a nerve for a few Vermont public servants. Two former state employees shared in the article's comments section candid stories that expose the "healthy tension" between commerce and regulation as a devastating barrier to state government's capacity to protect the environment. Bruce Post writes:

> When Gov. Richard Snelling died, Howard Dean, the new Governor, met with Snelling's staff in the fifth floor conference room in Montpelier's Pavilion Office Building. He was compassionate and very comforting to a staff that had just lost its leader, whom many of us respected and loved. (As Snelling's planning director, I was there.) Toward the end of his meeting with us, Howard Dean said, "My first priority is jobs, jobs, jobs."

I have reflected on that statement ever since, especially now as I research, write and lecture on Vermont's environmental history. Generally speaking, it seems to me that Vermont politicians, like those in other states, are more than willing to sacrifice the environment on the altar of "jobs, jobs, jobs." From Bennington in the south to Jay Peak in the north and many points in between, we may talk a good game about our environment, but in the end, Vermont's ecological health frequently takes a back seat to the priorities of commercial enterprise.

Stephanie Kaplan replied:

Bruce you're exactly right. The State's support for ChemFab despite the harm to nearby residents is consistent with the long-standing attitude on the part of various agencies in Vermont state government to turn a blind eye or even encourage corporate and especially industrial activity even if it may—and does—harm the environment or people who live in the area. I remember when I was working as an assistant attorney general in the 1980s and it was discovered by one employee of the Agency of Natural Resources that the Staco thermometer plant in Poultney was causing mercury contamination of the environment and the plant's employees who were inadvertently bringing mercury home in their clothes and contaminating themselves and their families. When I met with the state people in charge, one of them casually remarked that he had known about the contamination. When I asked him why he hadn't said anything he said—I quote his exact words—"I didn't want to open a can of worms." His implication, of course, was that he knew that his bosses would not want to have to take an action against a company in Vermont that CREATED JOBS for Vermonters. Never mind that these people with the jobs were being poisoned by the irresponsible and illegal practices of their employer and that this was known by an employee of the state.

Taken in sum, these stories reveal that even in America's bastions of progressive environmental politics, placating employers remains the least complicated and most popular approach to maintaining elected office.

In looking for evidence of reflexive ecological policies through the "Teflon Town" case, it would seem that even progressive officials become hardened because of the particular way they understand their role in Geels's web of "mutual dependencies" between government and private enterprise.[19] Former governor, physician, and presidential candidate Howard Dean is quoted in the news series as saying, "That's what government does. You listen to both sides and make a decision." Reading this statement, I was struck by how similar it seemed to a statement made by the PUC official with whom I had spoken about wind turbines. Both seem to articulate a political ethos that reduces the officials' job to brokering between conflicting interests. Democratic ideals and scientific evidence, it seems, are applied at the officials' discretion, and two sides to an argument are treated as equivalent despite gross disparities in power and influence. One can observe this ethos in practice when wind complaints are adjudicated. Mothers advocating for a quiet night's sleep for their children receive the same consideration as a developer securing profit for eager shareholders. One casts a single ballot in the voting booth, but the other pays for the political campaign and generates spending that sustains the economy or keeps scores of voters employed.

We see from the "Teflon Town" case that by seeing themselves as anointed decision-makers, public officials deploy ecological reflexivity only to the extent that it suits their particular position. For example, Howard Dean regrets that the pollution's journey from the smokestack to groundwater took decades. Dean surmised in his newspaper interview that water contamination at the time would have been more vigilantly policed because high-profile groundwater scandals had already raised public concern, meaning that regulators would have had popular support to intervene. Ostensibly, he means that regulators could have aggressively enforced environmental laws against a major employer without hurting Dean's prospects for reelection, because the public would have been on his side.

Dean's reflection on how officials need political cover to rein in polluters is revealing of inequality's cost to the environment. In a society with few social safety nets and a wide segment of the population living on the edge of financial abyss, rural jobs are more important than enforcing environmental laws, not just because companies are powerful, but also because half of the public's priorities are forged in relative deprivation. No politician can afford to kill rural jobs when rural jobs are few, which brings light to another way that the joint causes of social and ecological exploitation can and should be recognized.

Conclusion

Weighing the views of opposing theoretical camps forces us to look at an issue from multiple perspectives. Exploring how the ToP and EMT perspectives differently frame wind energy in Northern New England yields conflicting evidence. Vermont's advancements toward ecologically modern policy goals are impressive and extensive relative to the progress of other states, but Vermont's advancements have evolved as part of treadmill politics.

Like many Vermonters, I want to believe that our tiny state with a bleeding green heart is advancing cutting-edge environmental legislation, but in distinguishing between rhetorical and empirical progressivism, a gaping chasm appears. Over decades, leaders have extolled their commitment to natural resources but they have failed to practice ecological governance of the most basic sort, which should engage with science and prioritize sustainability and human health. Environmental scandals in Vermont now seem to bubble up regularly. After Teflon Town, it was regulators' renewal of a landfill's operating permit, despite evidence that it leeches toxins into a pristine international lake supplying drinking water for over 175,000 Canadians, in which fish now suffer cancerous lesions at alarming rates.[20] Next came the revelation that nearly twenty thousand tons of glass that Burlington residents recycled between 2013 and 2018 had been dumped in landfills.[21] Nationally, recycling programs are in shambles because the market for many post-consumer materials has fallen out and because China stopped accepting our plastic refuse, much of which was never being recycled anyway. While communities around the country are ending their recycling programs, solid waste administrators in Vermont's most populous and progressive county quietly buried their glass rather than confront the green public. These stories should raise doubts about the depth of Vermont's environmental resolve. This track record suggests that green jobs came to Vermont because green people wanted to create them and fill them, not because green politicians were more committed to them than to any other kind of job. Similarly, wind energy offered leaders a politically popular and inexpensive form of economic activity supported enthusiastically by utilities pursuing profitable new generation assets.

It seems that every day we face new headlines about climate change: models predicting faster warming, permafrost reaching a tipping point, the Amazon losing its capacity to sequester carbon. In the United States, where the federal preference for neoliberal climate solutions presents a high barrier to emissions reductions, state legislators and governors have an important role to play in delivering meaningful policies and programs.[22] And yet one of America's most environmentally progressive states has an energy strategy wholly reliant on selling renewable energy certificates that facilitate emissions elsewhere—and that can exist at all only because fossil fuels remain integral to the national electricity grid. This type of policy fuels the treadmills of production and consumption, underwritten by non-renewable resources.

While it may be true that Vermont's close relationship with GMP helped set an aggressive renewables agenda, it is plausible that such a relationship means Vermont's agenda is more influenced by private profit motives than that of any other state in the region. This probably explains why a larger carbon footprint for Vermont ratepayers resulted when RECs were first traded to power plants in other states. Optimists might maintain that REC markets slowly incorporate ecological reflexivity into society's energy decisions, but in light of the dire extremes we face, we should optimize our energy options for their greenhouse gas reductions, not their profitability. When policies allow green electricity executives to throw their customers under the renewable energy bus, it raises questions about whose reflexivity and whose ecological identities matter in such an energy paradigm—and it certainly makes a weak case for ecological modernization.

9

Energy and "Justice" in the Mountains

••••••••••••••••••••

Wind opponents planned a "day of action" at Vermont's capitol in January 2016. The event coincided with testimony about energy siting before legislative committees in the morning and concluded with an afternoon press conference announcing a proposed bill to limit new wind projects. The building, which we call our State House, sits on a hillside that casts its granite facade and gold dome against a forest backdrop. Compared to other state capitols, the State House is small and accessible, the embodiment of open government. The atmosphere inside is usually affable if not downright folksy. Compared with other centers of power and status, one might be surprised by the open display of natural hair colors, sweaters, fleece vests, and clunky winter boots, among all genders.

I made my way inside after photographing the cobbled-together wind turbine stationed in protest on the State House lawn. In their staging area, the day's organizers had left behind three easels holding posters and maps tracking Vermont's "rebellion" against renewable energy. I learned that nearly a quarter of the state's municipalities, more than a third of its land area, and two-thirds of its legislative districts had joined the movement to mandate local involvement in wind turbine siting. Delayed by snow, I missed the

FIG 9.1 Wind turbine erected in protest on State House lawn, Montpelier, Vermont, January 2016 (Photo courtesy of Shaun A. Golding)

opening remarks, so I proceeded to poke my head into a few overflowing committee rooms to hear public testimony.

Like most press conferences at Vermont's State House, this one was held in the Cedar Creek Room, a large, pink-hued Victorian-style sitting room that functions as a hallway between the old and new wings of the building. The room is named for its larger-than-life Civil War painting depicting Vermont soldiers' heroics at the Battle of Cedar Creek. The speeches that day would draw from this military imagery in framing the fight for local involvement in energy siting.

As the event drew near, wind opponents clad in neon yellow safety vests filled the room. While the wind opponents were nearly 100 percent Caucasian and over fifty years old, it was clear from their clothing and manners of speaking that the group included people from diverse socioeconomic backgrounds. The vests showed the group's solidarity and drew attention to the human impacts of energy development and the industrial nature of grid-scale wind turbines. The vests also drew immediate attention to protesters' bodies, as this constituency has felt ignored by lawmakers and regulators. The group also distributed informational literature in packets of brightly colored paper, and, with their vests and their pamphlets, the protesters were loudly demanding attention.

The press conference was called by Senator John Rodgers, a Democrat who hails from the town of Essex in Vermont's least economically vibrant region, the Northeast Kingdom. In calling for change in the state's energy siting regulations and championing this day of action, Rodgers had quickly made friends among wind opponents. The room erupted in applause when he announced his intention to introduce legislation banning industrial-scale wind turbines, as if it were a surprise. No doubt, most of the audience learned of his intentions a month earlier from a newspaper editorial. Rodgers's vow to require that energy projects undergo the same environmental review as other development projects garnered an equally exuberant response. In his remarks he observed that Vermont's energy landscape had begun a rapid transition away from large, centralized generation, and that this change should have entailed a parallel transition in how Vermont regulates new electricity infrastructure.

To reiterate, Rodgers is a Democrat putting the brakes on renewable energy in perhaps the nation's most environmentally progressive state. To understand why this came to pass, one must consider the region's geopolitics. Rodgers represents the poorest, most rural part of the state, an area that has been targeted by "Big Wind" more than other areas. While some may believe that no development should be turned away, others scrutinize the few long-term jobs promised by renewable energy. With large solar arrays, critics take issue with the loss of forest and farmland, or space that could otherwise host sorely needed development to bolster rural economies and keep young people from leaving the region.

Don Cioffi took the podium next. Cioffi is an administrator from a town seeing aggressive solar development. Cioffi's community had seen several hundred acres of forest and farmland targeted for solar development. By some reports, so many big projects were exporting electrons from his region that the grid could not accommodate new rooftop solar installations. His community underwent a public planning process to address the new energy-related land uses and drafted codes to restrict wind and solar projects in their scale and location. The town's position was that its existing use of solar power offset the town's carbon footprint and that, in the face of continued development, the town needed to protect its prospects for other economic activity. According to Cioffi's remarks, the community's codes were "summarily rejected" by Vermont's Public Service Board, now called the PUC.

Cioffi spoke with populist fervor. Throughout, he likened wind developers to "the aristocracy" and his audience of demonstrators to "yeoman farmers." He invoked the imagery of soldiers, suggesting that wind opponents were rallying to overthrow an oppressive regime led by "out-of-state millionaires," which he called the "renewable industrial combine." Cioffi referenced the expression "let them eat cake" to underscore the widespread perspective that energy decisions that are central to ordinary residents' lives are the domain of a few elite entities with little understanding of local perspectives.

I left the press conference uneasy. Of course I see the exploitative power relations in the region's wind industry. But something about the dramatized rhetoric of exploitation was not sitting well with me. That morning I had walked past a single tattered snow-covered "Black Lives Matter" sign but a multitude of antiwind signs. It was the same day that the White House declared the water crisis in Flint, Michigan, an eighteen-month poisoning at the hands of bumbling, criminally indifferent bureaucrats, a "federal emergency." The antiwind rally's crowd outnumbered by twenty times the audience for an earlier press conference for programs that give rural schools access to healthy, local food. I did not feel as though many wind opponents saw their fight as related to all the other struggles going on in the world, and it did not seem that wind opponents cared about these other struggles nearly as much.

Naomi Klein's 2015 book, *This Changes Everything*, presents climate change as a crisis born from injustice.[1] Klein links Western society's treatment of vulnerable peoples to its treatment of the planet, noting that our institutions reduce both humans and nature to business inputs, or "factors of production," as sociologists say. She argues that saving the environment cannot occur without acknowledging, healing, and empowering the peoples and cultures also exploited by the pursuit of profit. Understood in this way, the climate crisis is a challenge of numbers as well as morals. While environmental movements seek more members, those members must resist the temptation to put their ecological concerns over the unmet needs of fellow humans struggling for basic rights.

Americans pursuing renewable energy would be well served to make social justice a central cause in their agenda—and I mean global social justice, not simply justice for wind neighbors. Energy is, after all, a central character in the story of capitalism's conjoint exploitation of humans and the environment. Human energy justified feudalism, enslavement, and the invention of racial categories. Then, the energy in hydrocarbons and electric wires made some humans less essential to the economy as workers but more essential as consumers. Meanwhile, networks of extraction and movement allowed energy capitalists to hide their dirtiest and most inhumane practices from their consumers while amassing enormous wealth. New energy systems will move beyond this pattern only if society demands. But so far on the U.S. energy transition, workers seem more precarious and customers further in the dark about the consequences of their consumption habits.

This chapter tries to illuminate some of the darkness and ethical gray areas pervading pro- and antiwind conversations. The chapter problematizes the way wind resisters and wind supporters reference energy justice in strategic and self-interested tropes, focusing on place, race, and inequality.

What Is Environmental Justice?

Social scientists understand environmental injustice as an unequal distribution of ecological burden. It is a primary focus of Environmental Sociology and decades of study show that throughout the world, it is primarily poor, indigenous, and nonwhite groups who live and work in close proximity to pollution and toxic conditions. These groups experience disproportionate rates of sickness and disease, and these groups die younger than wealthier, White groups. The best studied examples of environmental injustice focus on industries that expose workers and neighbors to chemicals, some of which accumulate over regular exposure, such as the Teflon case from chapter 8. Nationally prominent cases include Louisiana's high density petrochemical "Cancer Alley" and the toxic chemical seepage portrayed in the film *Erin Brockovich*. Globally prominent cases include Shell Oil's ecological devastation in Nigeria, Amoco's activities in Ecuador, and Dole's poisoning of its banana plantation workers as portrayed in the 2009 film *Bananas*. But really, the examples are too numerous to count, even those documented in film: slowly leaching landfills and manure lagoons; lead in soil, paint in rental housing, and municipal water; particulate matter in urban air; toxins in foods relied on by society's most vulnerable. In most cases, those with social agency are able to avoid, minimize, or remedy their exposure, while those who lack agency must live their lives under elevated exposure, often though not always knowingly. Electricity production from radioactive and fossil fuels leaves an especially enduring mark on the poor, with exposure to coal and uranium mining, refineries, pipelines, fracking wells, and air pollution disproportionately impacting inner cities and rural areas.

Even in affluent countries like the United States, people live in toxic hot spots because the regulatory and enforcement capacity meant to protect citizens from pollution is insufficient. However, the reason toxic hot spots exist in the first place is a function of the way our society makes sense of and practices justice. Put simply, we operate under a notion of justice attributed to philosopher John Rawls. Rawlsian justice relies on the assumption that goods are distributed so that the benefit of many can justify the harm of a few, so long as those few are still better off than they would have been under an alternative scenario.[2] In other words, we accept pollution because polluting industries advance the overall level of wealth and comfort felt within the country at large. Those who live downwind from the smokestack at least benefit from the electricity generated in the power plant, we tell ourselves. Though Americans enjoy relatively strong legal protections for clean air and water, those protections exist within a broader political paradigm that justifies the sacrifice of few for the betterment of many.

We know this mindset is problematic. To what extent does one benefit from industrialization if one is hospitalized, ridden with tumors, or dead by forty? This is why attention to environmental justice was codified by the Environmental Protection Agency in the 1990s. Nonetheless, in spite of formal attention by federal mandates, environmental justice remains a social problem with costs that can be tallied in terms of disproportionately high hospitalization rates and cancer diagnoses, and lower life expectancies.

Simply recognizing the existence of environmental injustice has done little to eliminate it. Both activists and scholars have made the compelling case that environmental injustices derive not only from systematic exposure to hazards, but from the categorical exclusion of poor and minority groups from society's basic economic structures as well as the regulatory apparatuses governing environmental health and safety.[3] In other words, intense social and economic inequalities, and not just dirty industries or indifferent bureaucracies, put environmental justice out of reach.

Politically progressive thinkers have been conditioned to believe that wind turbines embody environmental justice, and yet debates around renewable energy reveal extensive ambiguity about energy and fairness. In pursuing energy justice for themselves, Northern New Englanders talk a lot about injustices on others' behalf, but seem to have difficulty allying in a way that actually promotes a just energy system. This shortcoming is the product of a bewildering energy landscape and conflicting worldviews that spin off self-serving assessments of fairness, which I propose are rooted in the region's history and environmental culture.

Wind and Injustice: It's Complicated

In throwing arguments like pasta, wind opponents occasionally invoke narratives of justice, out of apparent desperation to win over the public. The particular ways wind opponents do so reveal a great deal about the growing gulf between social and environmental justice as it relates to clean energy.

Energy scholars Sovacool and Dworkin outline three priorities for energy justice that resonate with the broader tenets of environmental justice.[4] The authors recognize that energy's costs—the ways the hazards and externalities of the energy system are imposed on communities—are often borne unequally. Similarly, energy systems' benefits—the ways members of society access and use modern energy systems—are often highly uneven. Lastly, the procedures through which energy projects proceed through regulatory approval often involve exclusion from decision-making.

Against these benchmarks, Northern New England's wind opponents feel like victims of energy injustice. Their share of wind energy's costs and benefits is brokered by complex policies and external interests profiting from the region. Additionally, they are excluded from meaningfully participating in regulating existing turbine installations, much less intervening in the proposal of future turbines or charting energy priorities more broadly. Contributors to public forums and my informants have used the phrases "social justice" and "environmental justice" to describe the perceived inequity of siting wind turbines in poor rural communities, as have energy scholars.[5] But is this what Naomi Klein was talking about? Should we think of energy justice as an extension of social justice, environmental justice, or both?

While it is undeniable that energy siting, including siting of renewables, often saddles the most vulnerable with the harshest outcomes, my observations in Northern New England reveal the need for caution before equating every case of renewables siting to a breach of social and environmental justice. Starting with sociologists' conventional notion of justice, ridgeline wind perfectly illustrates Rawlsian justice because siting undesirable infrastructure in sparsely populated places reduces or contains the negative impacts of a development from most people, who in North America now live in cities and suburbs. Under the Rawlsian paradigm, wind turbines on rural ridgelines are the embodiment of fairness. But several wind opponents believe that communities hosting turbines are selected because they are poor, which presents a different lens for assessing fairness.

Indeed, many turbines are built in communities experiencing postindustrial decline. They have low property values, high poverty and unemployment rates, and accelerated aging and out-migration. Turbine communities are thus the "path of least resistance" for developers. Not only do these communities have fewer resources to mount legal opposition, but in several cases they are also enticed by the promise of jobs, infrastructure, or tax relief. Because it is unclear whether host communities are selected for their poverty or their population density, the implications for justice are murky as well. Siting turbines where neighbors have the least power to resist them exploits inequality, and both society and social scientists tend to regard that very differently than minimizing risk. In other words, it may seem fair if the rural poor are subjected to undesirable things because they are rural, but it is different if this happens to them because they are poor. Rural Northern New England has enough pockets of affluence conspicuously devoid of wind development for observers to speculate that locations are indeed selected for their incomes and not necessarily their low population density.

Even still, are wind turbines as problematic as inner-city gas power plants? Should we advocate for a different framework of justice in wind turbine siting? Again, the answers are complicated. It remains a legitimate albeit unpopular perspective that wind turbines make terrible neighbors, and in fact, court rulings and government directives are acknowledging this perspective. A high-profile case in Massachusetts resulted in the early demise of a municipally owned wind turbine because of neighbor complaints.[6] Even in Northern New England, turbine siting standards support the legitimacy of claims often dismissed as selfishness. Turbine noise standards identify a decibel level meant to establish what constitutes an equitable shouldering of society's energy burden. Households in which turbine decibel levels exceed those standards have a legal right to identify as victims of an unjust energy hosting arrangement. Thus, it is false to assume that the renewable energy transition is immune to critiques of its fairness. As prominent cases have revealed, it is poor families living adjacent to wind turbines who have the fewest options to remedy complaints about noise and shadow flicker, much less measure and enforce state-mandated standards.

Amid this complexity, talking about justice is also complicated. While it is certainly possible to identify scenarios in which Northern New England's wind industry has failed to provide energy justice, many of those engaged in debating wind invoke justice with wide latitude. The fight against wind turbines raises several questions: Is it just for wind supporters and wind opponents to draw parallels with the environmental justice movement when few similarities exist? Is it exploitative to articulate injustice on communities' behalf without consulting or assisting those communities? Who gets left out altogether when communities in affluent countries debate renewable energy justice?

Justice by Proxy at Three Scales

Those Poor People: Place, Class, and Victimhood in Northern New England. As suggested in its name, New England is anchored culturally in the rigid English class system. This anchoring is embodied in the region's centuries-old elite institutions, resort enclaves, and displays of cultural capital. For example, small talk among strangers often begins with queries about one's occupation, and sometimes one's parents' occupation as well, because such information conveys fodder for categorization that the regional culture values. (It wasn't until I lived in other parts of the United States that I realized how elitist this seemingly innocuous habit is.) "Summer" is used as a verb by many, but strictly as a noun by many more. In rural Northern New England, high school hockey and lacrosse teams tend to signify the communities with Ivy League parents. Rigid stratification can be measured empirically in home values, income, employment, and college attendance among high school graduates. And now, even in the most remote corners of the region, hybrid cars and residential solar arrays overlap with more established status symbols. Whereas affluent mountains towns have ski trails, poor mountain towns have wind turbines.

Wind opponents feel their region is being exploited, but many with whom I spoke realize that they personally enjoy relatively high social status and instead make claims of injustice on behalf of disadvantaged communities. They argue that the poor, sparsely populated communities where developers site wind projects were targeted strategically to exploit economic desperation.

This characterization is problematic because wind turbines' impact on the financial well-being of communities cannot be reduced to one archetype. Direct payments to municipalities, which in some cases completely erase households' municipal property tax bills, help keep poor landowners on their land and buoy school budgets that might otherwise wither away due to declining enrollments. On the other hand, the big money rarely goes to locals. While owning the land on which turbines are built allows landowners to secure long-term payments or royalties, in Northern New England, that privilege extends primarily to absentee landowners who own entire mountains. Few farmers own land suitable for industrial wind development, and those who do have met with heavy opposition.[7]

Regardless, against the region's patchwork of social classes, claims of environmental injustice by proxy ring hollow when communities debated the issue openly and voted to accept payments to compensate them for the arrangement. And yet for some opponents, communities that voted in favor of ridgeline wind were even more reviled than developers who targeted those communities. Ralph discussed one community's vote as tantamount to committing welfare fraud, as if being abandoned by shuttered factories and mines made those communities prone to finding lazy solutions to their needs, saying "the first opportunity that they have to sell out to a big company and sell the mountain, they did it, and that makes me furious." Jim has a similar take: "They began to look for the next big. . . . It's like somebody playing the lottery and their lottery ticket came in with this project." In a reflection of society's frequent shaming of the poor, communities that accept wind are talked about as both easily exploitable and excessively willing to be exploited.

While opposition to wind development has united individuals across very diverse income levels and occupations, application of the term "justice" to wind turbine siting illustrates the deep divides between the region's rich and poor municipalities. According to many opponents' logic, if there were no poor communities, there might be no turbines on the region's mountains. So, one might ask, were wind opponents aware of inequality before the wind turbines brought it to the opponents' attention? How, if at all, did they show concern for underdevelopment in the region's poor communities before wind turbines threatened the landscape? While Vermont's Senator Bernie Sanders has worked tirelessly to raise awareness of inequality while running for president, the state he represents has shown no groundswell of support for the issue on par with its environmentalism, even as whole areas experience gutted real estate values and population loss.

At several moments over the course of my research, I was struck by the region's inequality and how nuanced social class became as it manifested around the wind issue. One moment in particular stands out. At a public listening session, over a proposed ridgeline development in Canton, Maine, one could sort participants into class categories by their cars, clothes, and accents, but those categories did not reveal anything about how the participants felt about wind turbines. This ambiguity turned my first wind meeting into a guessing game of sorts. I was, however, completely blindsided by one citizen's testimony.

In her sixties, with short, graying, curly hair and strong tanned arms, a woman took the microphone in front of the state's Public Utilities Commission, who by that time looked bored and tired in the sweltering, crowded fire hall. The woman identified herself as a widow and a cancer survivor. She had spent her life savings on a piece of land and a mobile home abutting the site of the proposed turbines. She testified that she had been slowly saving the funds to purchase a septic tank and drill a well for her home, and that in the interim she used an outhouse and carried water to her home in jugs. She testified that without the plumbing, which she could not yet afford, Maine's energy siting law did not recognize her as a legal neighbor, and thus, it did not recognize her right to demand turbine setbacks from her domicile or limitations on sound penetrating her walls. Vermont's siting law uses the same standards, which has meant that poor or intentionally low-impact neighbors have little say in large developments or claim to mitigations. In this case, because she was too poor to equip her residence with electricity or running water, she was not considered worthy of a voice in the process. Her story puts into sharp contrast the difference between environmental decisions made out of necessity and those born from privilege. Living on Maine's spatial and economic margins, she would shoulder the greatest burden of renewable energy but receive the fewest benefits. Cases like hers are exceptional, but no onslaught of legal support for those in her position has emerged, and no change in policy has resulted.

As I learned in Canton, Maine, there are egregious barriers faced by poor neighbors to wind facilities, and given the profits to be made by large wind installations it seems especially cruel for developers to skimp on buyouts. But, as part of the antiwind strategy of throwing pasta to see what sticks, projecting victimhood by proxy onto entire municipalities is to use one's neighbors instrumentally without the courtesy of consulting them, much less helping them directly. Moreover, using "social justice" or "environmental justice" in relation to wind development seems especially disingenuous relative to the imminently toxic threats saddled by poor and nonwhite communities without compensation.

Those Poor Indians in Canada. Wind opponents have not cornered the market on justice by proxy rhetoric. Wind supporters use it too when they reference Quebec's indigenous peoples impacted by hydroelectric development.

Through ownership of its hydroelectric utility, Hydro Québec, the province of Quebec commenced its first major dam projects with little regard for the cultural dignity or treaty rights of indigenous groups. The utility's dams submerged ancestral homelands and hunting lands and decimated productive fisheries. The dams changed the seasonal flows of rivers and halted millennium-old subsistence patterns for several First Nations communities, who have seen their health and livelihoods deteriorate as a result.

Even tribal communities that eventually negotiated settlements from Hydro Québec show the cultural wear of colonialism that comes with the intertwined decay of centuries-old ecologies and livelihoods. In those communities, well-maintained public facilities serve a population battling joblessness, homelessness, and substance abuse, inspiring many to decry the continued development of hydro projects on the grounds of cultural genocide. Damming rivers has also had medical consequences because of toxic accumulation in reservoirs. A recent story focusing on mercury poisoning passed to humans through fish included a message from a Canadian man to electricity customers in the U.S. Northeast: "You're buying the misery from the local people of northern Canada. That's not a good thing."

These news stories offer wind supporters moral ammunition. References to Quebec's indigenous groups are occasional add-ons to the "we're doing our part," argument supporting wind development, which people make when others cite hydropower from Quebec as preferable to wind. In effect, wind proponents position local wind energy as being morally superior to electricity purchased from Quebec.

While the plight of Quebec's First Nations certainly constitutes a severe and enduring injustice, the way it enters New England's energy debate is similar to facile claims by proxy leveled by wind opponents on behalf of turbine host communities. Opponents making these arguments instrumentally project victimhood without so much as consulting the victims, much less assessing or addressing the unjust conditions they experience. The New England grid has purchased energy from Hydro Québec for decades, but even among the region's emerging energy activists, I encountered no inkling of a plan to remedy the injustices experienced in Quebec's First Nations.

Those who reference Quebec's indigenous groups are of course trying to make the point that even "clean" energy comes with consequences and that for New Englanders to show more concern for its mountaintops than for human beings is morally suspect. This is of a course a valid point, but one that serves the personal conscience of American energy consumers far more than the human beings in question. Rather than address the injustices of Quebec's hydropower, ridgeline wind projects allow ratepayers to dissociate themselves from and absolve themselves of responsibility for these injustices, making indigenous groups little more than rhetorical props in a conversation to which they should be central. Buying less hydropower from Quebec will not halt dam construction, so using Indigenous Canadians in pro-wind arguments holds no other logical purpose than to add moral superiority to the platform.

In the state of New York, there is a precedent for collaborating with Canada's indigenous people actively rather than referencing them passively. In the 1990s, with Quebec poised to build new dams with capacity to sell electrons south of the border, the Cree Nation launched a publicity tour to raise awareness of its injustices. Paddling by canoe to New York City, Cree dignitaries convinced New York governor Mario Cuomo to cancel a twenty-year contract for hydroelectricity. With this leverage, the Cree eventually brokered far better financial terms for their sacrifices.

Thirty years later, New York's collaboration with Quebec's indigenous communities seems to be keeping pace with liberal voters' growing intolerance for injustice. With newer, larger hydro projects in Quebec readying to deliver electricity through a direct cable to New York City, the mayor's office has proactively and directly engaged with Quebec's indigenous groups in advance of final negotiations with Hydro Québec. The mayor's director of sustainability told the *Montreal Gazette*, "We need to make sure any deal we become a part of is consistent with our values and one of those means that we are constantly looking at the fight against climate change and our fights against inequality to be one and the same." A tribal leader told the newspaper, "It's the first time in my years of leadership that I see such a just and communicative approach, especially coming from people in the United States."[8] While it is too early to tell whether these events will lead to substantive improvements in the lives of Quebec's First Nations, it sketches an outline for a model of what collaborative and just energy sourcing can look like.

The story in New England has yet to evolve in this direction. With nuclear facilities aging out, and the cost of replacing them skyrocketing, progressive cities of the American Northeast have few low-carbon electricity options. Hydro Québec plans to dam several new river systems to accommodate its southern neighbors' electricity demands, and current debate is about how to route power from Quebec to southern New England's cities. From this corridor debate, a new critique of Hydro Québec's treatment of an indigenous group emerged. In the 1950s, the Pessamit were the first of Quebec's tribes to have its tribal territory flooded. The Pessamit received all of the cultural and ecological hardships later endured by the Cree, but with a tiny fraction of the compensation. In 2017, the Pessamit sought to raise attention to this injustice by weaving their story into the Northern Pass debate, in which New Hampshire was fighting to prevent its prized White Mountains from becoming an energy corridor for Massachusetts.

News stories across New England, including a four-part feature from New Hampshire Public Radio, told the Pessamit's story to Americans in great detail.[9] Like the Cree in the 1990s, the Pessamit hoped their PR campaign would garner enough support from the American public to persuade Hydro Québec to address the tribe's claims. In particular, the Pessamit sought acknowledgment of the disastrous impacts Hydro Québec's dams have had on the Betsiamites River's wild salmon stocks.

There is little evidence that transborder solidarity is aligning to help the Pessamit as it helped the Cree. When New Hampshire's siting commission ruled against the Northern Pass, the action scuttled the Pessamit's media attention. With Massachusetts utilities weighing new corridor routes, that state's progressive leaders and voters could commit to advocating on behalf of the Pessamit when negotiating power contracts, but that does not appear to be on anyone's agenda. Hydro Quebec has refuted many of the Pessamit's claims with jargon-laden talk of its indigenous partnerships and dubious ecological successes. Despite their best efforts, the Pessamit remain a footnote in Northern New England's energy conversations. In the news story on mercury poisoning described earlier, one New Hampshire legislator told the reporter, "I say, to the First Nations people up in Canada, I say 'go fight that. It's your right to stop it if it is harmful,'" he said. "I think they should fight it. But it's not my fight. I'm not a citizen of Canada, so I can't be fighting elsewhere in the world."[10] This stance treats the U.S.-Canada border as if electrons, money, and humanity pass across the border differently. Such a perspective syncs in lockstep with wind proponents' preference for local electricity that allows them to feel morally superior without having to actually do anything specific for Canada's First Nations.

Climate Change and Global Justice. In Northern New England's wind debate, for all the talk of fairness on others' behalf, the planet's most vulnerable people would appear to have no voice. The term "climate justice" acknowledges that the world's poor will suffer the most widespread and enduring impacts of global climate change. Inhabitants of low-lying islands are already losing their homes. Climate models predict weather patterns conducive to pests that will spread disease across human, plant, and animal populations and disrupt agriculture. The capacity to survive these hazards lay with those who have wealth, land, and access to health care.

Though we think of renewable energy as a means of mitigating hazards, we put very little effort into verifying, much less ensuring it. Recall how when pressed, Vermont bureaucrats so easily deferred to cost savings, not emissions reductions, in justifying wind development. But addressing climate justice requires more than reducing our climate impacts, it requires doing so through just means. We assume that transitioning to renewable electricity *is* justice, but like most industries, renewable energy relies on resource extraction and manufacturing in far reaches of the planet, sometimes under tyrannical labor practices and political regimes. Yet, in discussions of energy justice in affluent nations, global social justice fades quietly to the background. Despite clear connections to vulnerable communities through the electricity grid, supply chains, and consumption habits, Northern New Englanders debate renewable energy with little to no regard for those communities.

World Bank estimates cited by the *Wall Street Journal* predict that converting to 100 percent renewables will require 34 million metric tons of copper, 40 million tons of lead, 50 million tons of zinc, 162 million tons of aluminum, and no less than 4.8 billion tons of iron.[11] Currently, mining and processing these materials and assembling them into turbine parts releases toxic emissions impacting rural communities in Africa, Asia, and South America. The same *Wall Street Journal* article notes that "mining has become one of the biggest single drivers of deforestation, ecosystem collapse, and biodiversity loss around the world. Ecologists estimate that even at present rates of global material use, we are overshooting sustainable levels by 82 percent." Moreover, in 2020, more than 80 percent of the U.S. supply of "rare earth" minerals, the most hard-to-find but essential battery and renewables components, were sourced in China,[12] a country suspected of forcing political prisoners to work in its mines.[13]

At no point in my research have I encountered discussions of the human impacts of mining and processing minerals used in turbines. Antiwind rhetoric treats turbines as toxic structures. Implicitly, toxicity is understood as a local ecological threat, never acknowledging the implications for those who mine and manufacture the materials used in the turbines and only rarely acknowledging the ecological devastation wrought by mining. It is most common for wind opponents to assail wind developers for failing to disclose the hazards and risks in each turbine, which include hundreds of gallons of petroleum-based lubricant. Thus, turbines are talked about as secretly dirty but not as secretly exploitative. I have encountered no advocacy for mineral sourcing standards that would protect the health and safety of overseas workers. Similarly, turbines, like any electronics technologies, will one day cease to function and require recycling to ensure that their toxic innards steer clear of human bodies and water supplies. This recycling happens primarily in poor countries where electronics are dismantled and sorted by hand with no oversight ensuring that health precautions are taken. Antiwind rhetoric emphasizes the high costs of decommissioning retired turbines as a threat to ridgelines. Wind opponents worry that the turbines will rust in place instead of being dismantled and removed. To my knowledge antiwind activists have generally ignored the high probability that any recycling that is possible will be outsourced to a poor country.

Many of the same production life-cycle critiques apply to solar infrastructure, which is being produced in exponentially increasing quantities and which wind opponents tend to favor. Solar advocates, however, seem unconcerned with how and why solar panels can be made so cheaply or what happens when the panels reach the end of their working life. Estimates suggest that in twenty-five years the current boom of solar electricity will result in twenty-three tons of additional e-waste. While the European Union has enforced recycling plans for new solar components, currently there are no similar requirements in the United States, much less broad standards of just sourcing and recycling in the renewable energy industry.[14] As Environmental Scholar Dustin Mulvaney argues, the problem is not that solar energy is impossibly dirty or that we should stop pursuing it, the problem is that conventions in the industry create victims, and that as the industry rapidly expands so too will the number of victims.[15] The obvious solution is to legislate standards of justice and sustainability into our renewables policies. While undoubtedly such standards would be popular in Northern New England, I have not encountered any interest in such standards in the region's renewable energy debates.

Renewable energy systems predicated upon "clean" technologies manufactured in poor countries exemplify an increasingly common dimension of climate injustice related to equity in both rhetoric and carbon accounting. Green technologies allow carbon reductions at the expense of poor nations, both in terms of local pollution and carbon emission allotments. In other words, the renewable transition in affluent nations is facilitated by delay in the renewable transition in poor nations. This moral oversight of accounting is referred to as "environmental load displacement" and "ecologically unequal exchange," and it is astonishingly pervasive in the way affluent nations celebrate environmental achievements as somehow good for the planet when they are simply displacing emissions and resource depletion.[16] Can rich countries celebrate their climate accomplishments if residents stock their stores and homes with products that have enormous carbon footprints counted elsewhere? Should rich countries continue to demand climate concessions from poor countries when so much of their greenhouse gas emissions go to satiating the rich countries' consumer lifestyles?

Thinking globally while acting locally is a resonant moral tenet in Northern New England's energy policies, particularly in Vermont's aggressive overhaul of its electricity generation. But who, precisely, groups are thinking about seems pretty clear as they invoke the concept of environmental justice without actually elevating the voices of communities marginalized by modern energy systems. It would seem that those debating renewable energy in New England invoke justice by proxy when marginalized groups have familiar names and faces, but ignore that renewables leave a diffuse trail of industrial pollution in developing countries, even as they cite health and safety concerns locally. Ultimately, these rhetorical tactics, along with the ease of mobilization for and against wind turbines, illustrate how much agency and political power most rural New Englanders enjoy relative to others.

The Racial Indifference of Energy Politics

When environmentalists battle big money, both sides are usually green and white—which poses a problem for aligning social justice and ecological sustainability in ways that are movement-building. How is a racially divided society supposed to perceive White people protesting assaults on the landscape when minorities are disproportionately incarcerated, murdered by police officers, and poisoned by environmental toxins? In Northern New England's wind debate one need not look far to find evidence that the environmental movement is too rooted in White upper class values to resonate widely.

The racial politics born from Northern New England's relative isolation sets the stage for the region's tendency to be good at talking about injustice but bad at doing much about it. To be clear, however, with renewable energy the problem is not that stakeholders are White, but that their agenda serves environmental priorities deeply embedded in a White ecological worldview. The American environmental movement was founded by rich White men and it has historically focused on preserving and conserving resources. That so many who move to rural New England are White upper class environmentalists from cities puts them in league with a particular brand of environmentalism—one that sees specific types of nature and places as worthy of defending while ignoring other types of nature and places. This history is the DNA of the way that many very committed environmentalists see, know, and care about the planet. But in an era now defined by globally interconnected and empowered postcolonial environmental movements, justice is more than a box to check in an environmental platform. Work has to be done to reimagine the platform with justice at its core.

Whiteness and Renewables up Close. In a homogenous social landscape, whiteness may never pose a problem to achieving local environmental goals; but whiteness came to the fore when wind resisters tried to build bridges to orient themselves more solidly in the camp of environmental justice movements. To their credit, an antiwind group recognized the need to link their resistance to issues of energy justice more universally, and so they planned a weekend of outreach and education one summer weekend. I will refer to the event as "Energy-Fest."

I was unable to attend Energy-Fest, but I followed up with one of its organizers afterward. My informant reported that the event attracted several hundred participants to workshops, public events, and two keynote speeches, one of which focused on the plight of indigenous groups. As Naomi Klein argues convincingly in *This Changes Everything*, indigenous movements are positioned to offer strong legal legitimacy to climate activism. Additionally, given Northern New England's long reliance on hydropower from Quebec, it made sense to include indigenous perspectives in the weekend's events. However, it seems that some of the organizers were unprepared to reflect critically on how their own perspectives embody colonialism.

The speaker identified as a Native North American. As the incident was recounted to me, on arrival, the speaker observed that the weekend's program included a workshop about ecological carrying capacity—the earth's ability to sustain human populations—and so the speaker immediately gave the organizers an ultimatum: cancel that workshop, or find another speaker.

The concept of an ecological carrying capacity applying to human populations is strongly contested and holds a bitter connotation for indigenous peoples. The premise of a human carrying capacity traces back to the end of the eighteenth century, when Thomas Malthus famously predicted that global civilization would decline because population was growing faster than food production. Malthus, an English aristocrat, attributed population growth to irresponsible procreation among the poor, whose sexual habits he found morally derelict and culturally inferior.[17] Malthus's predictions were laughably wrong, and yet a vocal contingent of primarily White, Western thinkers has continued to embrace the theory that population growth is imminently destroying the planet. Malthusian rhetoric is what is most reprehensible about Michael Moore's renewable energy film, *Planet of the Humans*.

Karen, my informant who has devoted much of her recent energy awakening to learning about fossil fuel victims in indigenous communities and poor countries, does not fit the Malthusian profile, but she confessed to me that she too has concerns about population. She confided, "This is so impossible to talk about, I think, without getting labeled, but I think we have a serious overpopulation problem globally. If we can't talk about that, we're screwed. We just can't. . . . I don't think the earth can sustain the number of people on it. That's one of the drivers of the economic machine that we're caught up in." The logical outcome of "talking about" overpopulation is excluding people from existing. I suspect that Karen understands this, and that is the reason she acknowledges the difficulty of talking about it.

We need only look at who talks about overpopulation and how they talk about it to better understand why it is a rhetoric fraught with racism. The reason Energy-Fest's speaker reacted so strongly is because they identify as a member of a group that has been victimized over centuries by Europeans' ideas about who should be allowed to exist. Today, White environmentalists routinely misattribute ecological problems to overpopulation, implicitly and sometimes explicitly blaming communities whose members tend to have large families. Researchers have encountered these sentiments in privileged White enclaves where environmentalism takes on racialized anti-immigration rhetoric.[18] Such rhetoric is obviously self-absolving among White Western thinkers, and it is a particularly stinging affront to indigenous communities, who stewarded diverse ecosystems for thousands of years and endured genocide, kidnappings, and forced sterilizations into the twentieth century.

Malthusian thinking is also intellectually myopic. Global environmental crises trace back to industrialization, which originated in affluent White countries that hoarded capital accumulated through slavery and colonialism. Global food systems certainly need repairing, but it is because they are primarily a means to further accumulation, not distribution.[19] And for nearly thirty years, social demographers have observed that lower fertility rates are most closely associated with women's education and labor force participation. So for those concerned about human survival on planet earth, breaking hegemonic control over food and promoting universal gender parity are platforms that serve your objectives without tacitly reproducing white supremacy.

The stand taken by the speaker at Energy-Fest exposes competing paradigms of environmental justice. Energy-Fest's organizers hoped to assemble a diverse program to draw out important connections between energy and inequity. But the organizers' conventional frames of environmental justice emphasize corporal victimization, harm to a human body in the name of profit. Western environmentalism is sometimes oblivious to other types of harm. For example, the way racist and colonial undertones pervade concepts like carrying capacity—the way its own platform perpetrates harm by silencing and excluding.

Entire cultures, ethnic identities, and ways of knowing suffer at the hands of well-intentioned but racist environmentalism. Describing how U.S. National Parks erased millennia of indigenous environmentalism in the name of White environmentalism, Native scholar Dina Gilio-Whitaker writes: "Born from the Manifest Destiny ideologies of western expansion, the preservation movement was deeply influenced by a national fixation on the imagined pre-Columbian pristine American wilderness and the social Darwinist values of white superiority . . . those legacies carried forth into twentieth-century environmental organizing...the result was a contentious—and sometimes openly antagonistic—relationship between modern environmentalists and American Indians, making the attainment of environmental justice for Native people more difficult."[20] In simple terms, our path-breaking American environmentalists held racist beliefs, and by sanctifying the experience of human-free wilderness, the American environmental movement perpetrated the cultural erasure and silencing of Native peoples from landscapes they were vital to sustaining. An event motivated by the loss of mountains that extends its arms to indigenous groups expecting solidarity replicates this problematic legacy of White environmentalism.[21] For the speaker at Energy-Fest, justice probably starts with acknowledging the harm and ecological devastation wrought from this history.

A more inclusive environmental justice requires centering all human communities as equally important forces in ecological maintenance and repair, not utilizing communities instrumentally. For example, traditional ecological knowledge amassed over centuries by indigenous societies has often been ignored or dismissed by scientific researchers, only to later be appropriated when validating evidence is "discovered."[22] Conversely, it has become common to romanticize and homogenize a pan-Native American perspective as a spiritual ecology, somehow in tune with natural rhythms.[23] Eco-spiritual language is often applied ambiguously to indigeneity, which reduces hundreds of distinct societies to a monolithic mythology, ignoring that the fight against genocidal Europeans may be the only widely shared trait among indigenous peoples.

"Settler colonialism" is projecting White European logics on every conceivable social domain in the process of erasing indigenous communities. It is replicated globally in policy and culture, and it is, of course, a counterintuitive solution for environmental problems. And yet, settler-colonialism is clearly reflected in industrial-scale wind development in poor countries, where European governments allow energy companies to outsource their carbon reduction commitments while ignoring local communities. Alarming examples are found in Oaxaca, Mexico, where European-owned wind turbines send electricity to Mexico City, bypassing local indigenous groups in terms of participation, compensation, and electricity.[24] It is clear, therefore, that there are parallels to observe in Northern New England's wind struggle and ongoing indigenous struggles, and yet the opportunities for unity and collaboration will be strained as long as a settler-colonial logic shapes environmental agendas.

The impasse at Energy-Fest illustrates this strain. My informant described the speaker's ultimatum and demeanor as "just not nice, very uncivilized," seeming particularly affronted that the speaker did not adhere to the group's cultural norms for etiquette. My informant told me, "Just because people are concerned about the environment and energy . . . you've got to be careful. . . . So, obviously inviting them was not the right thing to do. . . . And they didn't share our values about courtesy and politeness and, you know, not pulling in at nine o'clock and saying, 'Hey, I'm out of here.'"

Labeling an indigenous person or the person's behavior as "uncivilized" borrows from a historical script of bigotry. It was the perception of pathological incivility that settlers used to justify erasing indigenous peoples from their land in the first place, and of course it is unabashed effrontery for a White environmentalist to apply that label to a Native person when European *civilization* is precisely what degrades the earth's resources. (Incivility just might be the answer to all our ecological problems.) Further, to fall back on the cultural norms of "courtesy and politeness" is to saddle an indigenous person with the added burden of worrying about environmentalists' feelings, when the speakers' feelings seem to have been overlooked from the start. This is a problematic model for an environmental partnership.

Any fight for environmental justice that glosses over immense social inequities will struggle to enlist diverse support. Building true alliances requires decolonializing conventional approaches to conservation—for example, preventing the "natural" from taking precedence over the social, and perhaps reframing the very notion of the "conventional."[25] This extends to the ubiquitous environmental campaigns to save things like dolphins, campaigns that usually overlook the social-economic squeeze felt by locals who inadvertently kill the dolphins because their own survival depends on fishing. Acknowledging communities marginalized by White environmentalist norms is an important first step to making environmental movements more diverse.

The Mythology of Racial Progress, and the Green Policy Parallel. It is predictable when White environmentalism alienates people of color, but Northern New England and especially Vermont nurture a particular earnest but potent whiteness that is important for contextualizing the region's environmentalism.[26] In 2018, a *Saturday Night Live* sketch portrayed Vermont as Shangri-La for white supremacists because of its resurgent agrarian economy and lack of racial diversity. I found the sketch awkward on a number of levels. *SNL* has itself struggled for diversity in its own cast. (A handful of nonwhite cast members over forty years sort of makes it the Vermont of television shows, so naturally I saw some irony in the sketch.) What was most problematic was how the sketch erased Vermont's people of color and, in so doing, hinted at but largely glossed over the troubling ways that places like Northern New England engage with their very real, living and breathing minority groups.

Northern New England fosters a unique brand of racially insensitive politics, and the ease with which distant marginalized communities enter the wind energy debate as rhetorical instruments should be understood against the history of those politics. Though largely isolated from contentious racial issues, the region prides itself publically on vehement public stances about those issues. That is, we make many superficial attempts to distance ourselves from racism, but in doing so we fail to acknowledge that our regional culture reproduces a virulent strain of what sociologist Eduardo Bonilla-Silva calls "colorblind racism."[27] To illustrate, the region's basement crawl spaces and colonial-era root cellars are commonly mischaracterized as relics of the Underground Railroad, despite little evidence whatsoever in many cases.[28] Vermont proudly boasts its status as the first state to ban slavery in its constitution, but as historians note, an aversion to slavery is not equivalent to racial progressivism or even tolerance.[29] During the colonial era, several northern statesmen detested slavery because they feared slaves. They were afraid that slaves would rebel against their White captors.[30]

For most of the past two centuries, Northern New England's leaders were in no way immune to their eras' prevalent racial beliefs. All three states have an abysmal record dealing with their past and present indigenous peoples. Maine's eighteenth- and nineteenth-century shipping communities were pivotal to the slave trade, and Maine claimed a twentieth-century chapter of the Ku Klux Klan organized against French Canadian newcomers who labored in the state's textile mills. (These distinctions have earned Maine its own problematic portrayal on *SNL*.) Like many other states, Maine, New Hampshire, and Vermont each ran eugenics programs used to justify forced sterilizations into the 1960s. Eugenics programs, common across the United States, were state-funded initiatives to weed out racial minorities and other groups deemed bad for the human gene pool. This is the reason talking about "population control" should not sit well with anyone in America.

So the problem with the *SNL* skit and with regional environmentalism is the many ways that the region's mythology of racial tolerance becomes a self-congratulatory celebration, giving rise to a clueless brand of whiteness that minority groups in the region must contend with. In believing our regional narratives of racial progressivism and distancing ourselves culturally from overt racism, we ignore all the ways we overlook, underappreciate, and tokenize our nonwhite neighbors. Worse, in sustaining deep class divides, we fan the flames of status insecurity and white fragility, opening the door ever wider to overt racism.

Northern New England's whiteness is closely mirrored in *Saturday Night Live*'s whiteness. I wonder how nonwhite *SNL* cast members feel about being categorically excluded from a sketch meant to poke fun at racists? I wonder if anyone thought for a minute about how being a minority in an overwhelmingly white place must feel and how being overlooked on national television might not help all that much? To me, the sketch felt like having a party to celebrate progressive racial politics without inviting any people of color. After it aired, a Vermont shirt company sent *SNL*'s cast free flannels, and the state's community development secretary implored all visitors to Vermont to feel welcomed, even drumming up the state's "African American Heritage Trail."[31] But in toeing the "all press is good press" line with an upbeat sales pitch and defense of the state brand, those comments send the message that squeezing money out of tourists is more important than acknowledging that Vermonters of color do exist, or more important than rejecting white supremacism forcefully, which is sorely needed. Apparently, keeping Vermont's "African American Heritage Trail" markers from being vandalized by bigots has proved difficult.[32] After several highly publicized cases of racism in past years, Vermont's happy-go-lucky branding reads like whitewashing.[33]

I recount this story because environmental mythologies coexist with racial mythologies. Celebratory, self-congratulatory progressivism is an obstruction to a meaningful and just energy policy. Vermont, a state with no known petroleum formations underneath it, has banned fracking within its borders, but the state has done little to stem the flow of controversial fracked gas from elsewhere, despite intense public pressure. As some U.S. cities are banning new gas hookups, Vermont is defending a major gas line expansion. Maine communities have taken a hard line against allowing oil from the tar sands through a regional pipeline, but those same communities still fill their cars with gas refined from tar sands oil sent by train. Renewable energy policies also reflect this pattern. The region's per capita contribution to climate change is paltry and derived largely from transit and home heating, and yet, in turbines, politicians have chosen conspicuous and grandiose structures to advertise their commitment to carbon reduction. Vermont's weak renewables policy, premised upon the arbitrage of Renewable Energy Certificates, has made the very prominent infrastructure of renewable energy profitable without tackling greenhouse gas emissions as aggressively as other Northeast states.

Northern New England has a long history of looking at widespread societal problems from the outside. Distance, both geographic and cultural, buffers the region from contentious issues, allowing it to disengage from those issues substantively while addressing them with political platitudes. The region's ongoing energy debates, which juxtapose impassioned rhetoric with superficial results, has been shaped by this insular political history that so often relies on untested claims of progressivism. This history helps shape the region's pure and wholesome public image projected to outsiders, which becomes ingrained in individuals because it contributes to the larger narrative that we are principled people from principled states, "doing our part." But contrary to the progressivism symbolized by totemic icons of environmentalism, the region could be doing more for marginalized peoples everywhere—and more for the climate.

Conclusion

Wind energy in New England offers a window onto multiple nesting environmental injustices and their accompanying narratives, but distinguishing injustice in the case of wind development is complicated. The region's energy debate, as it unfolds in public meetings, hearings, and online forums, entails references to local injustice that rest on shaky ground and expose problematic inconsistencies in activists' pursuit of fair outcomes. In trying to claim a moral high ground, developers, opponents, and proponents alike reference marginalized communities and tell their stories as passive rhetorical strategies to advance the pro- or antiwind cause. Wind supporters and wind opponents both craft narratives of justice that reference marginalized communities without actually endeavoring to undo the conditions that marginalize them. On one side, wind is positioned as preferable to fossil fuels, an industry with a track record of treating entire communities as pawns, dumping grounds, and guinea pigs. The other side positions wind energy as the arm of big business, preying upon rural communities with no real accountability for keeping promises or even meeting climate goals. In both cases, these justice narratives are deployed as discursive strategies prioritizing local needs, with little regard for actually engaging with the plight of marginalized groups farther away. This rhetoric of injustice by proxy is uniquely tone deaf. Readers can probably appreciate the awkwardness of talking about someone in the room as if the person is not in the room. In modern energy politics, the whole world is in the room together, and yet we talk as if we are not.

Northern New England's resistance to wind turbines illustrates how well-intentioned environmental agendas can easily adopt elements of environmental racism. Self-interested and inward-focused policy among insular groups reinforces social barriers erected by legacies of colonialism. Northern New England's energy debate exemplifies this reinforcement of barriers, but a more widely relatable example is neighborhood segregation. Gross educational disparities have materialized as White households choose to live in homogenous neighborhoods. Those disparities widen immensely when White parents advocate for improvements that benefit their children's schools, instead of all schools. When electricity customers fight for or against wind turbines, are they thinking about the people who live next to coal power plants, or how those plants might provide the backup power for graceful new turbines? Desiring justly procured energy to satisfy one's own ethical conscience effectively cripples a more communal approach to just energy, one in which marginalized groups are included and not simply referenced. Erecting industrial wind turbines does little to guarantee that justice will be extended to those historically harmed by Hydro Québec or Shell Oil. Nor does wind opposition warrant the same label or remedy as the water crisis in Flint, Michigan, or the ecological and cultural devastation endured by Quebec's First Nations. The global environmental justice movement struggles daily to secure safe food, shelter, and work for society's most marginalized communities. It does these communities no good to appropriate the movement's label and apply it to White, landowning communities voting to host wind turbines.

Still, environmental injustices and energy injustices sometimes converge, and there is a need to more thoroughly explore this convergence in the larger context of inequality. Electricity's social and ecological impacts have always been unjustly distributed and we should not assume that renewable energy somehow erases this legacy. Energy's wide footprint of harm is the collateral damage of forceful and manifestly unjust global efforts to own and control a central factor of production. Social scientists studying renewable transitions should recognize that energy injustice exists in hegemonic energy systems around the world, and so it is elusive for many groups marginalized within their particular cultures and polities, and it often overlaps with environmental injustice. Energy justice is elusive for First Nations flooded from their homes and livelihoods, for coal miners left sick and jobless, and for poor neighbors to wind and solar facilities who pay more for their electricity than the urban households where the Renewable Energy Certificates are sold.

As Northern New England has shown us, it is politically fractious to ignore communities calling attention to these issues merely because their grievances are not as extreme as others'. This is particularly important as rural communities watch their livelihoods disappear and feel slighted by renewable energy policies devoid of equity. Society must openly deliberate the extent to which hosting infrastructures constitutes injustice, and the compensation and level of distribution that might be accepted as fair. If we allow those conversations to unfold freely, inclusively, publicly and across borders, we may bring to light the faults in our energy systems that marginalize people and the environment.

10

Reimagining Energy

• • • • • • • • • • • • • • • • • • • •

Renewable energy's need for open space is changing the footprint of electricity infrastructure, agitating place politics that have for generations excluded rural communities from deciding regional issues. This electricity transformation has come in tandem with quickening rural population loss and declines in prosperity. Thus, a growing number of rural communities in the world play host to conspicuous electricity apparatus of primary benefit to urban consumers, at a time when rural voices seem less relevant on the national stage. In this new rural landscape, wind turbines—for many, icons of environmentalism—loom above neighbors as daily reminders of their lost peace and quiet and their exclusion from political processes.

Northern New England's resistance to wind energy is not unique. Host communities that receive widespread compensation appear to experience the least antiwind discord. Neighbors who live very close to wind installations but receive little or no compensation appear to be most unhappy with wind turbines. These findings echo painfully simple recommendations often made to the wind industry: follow strong setback limits and offer more generous and widely distributed financial compensation for hosts and neighbors. Compensation, buyouts, and shared ownership would not have deflected the ire of those advocating for trees, bats, and mountains, but they could have potentially mitigated the intensity with which wind opponents in

Northern New England have disputed the science and financial accounting underlying industrial-scale wind. But then there are the issues of climate accountability and social justice to consider. Keeping local wind neighbors happy may do more for wind investors than for the planet.

Sociologists are not experts in grid-scale electrical engineering, thermodynamics, or climate science. We are, however, trained to evaluate the cultural, political, and economic influences of dominant environmental narratives, and we are capable of reflecting on how those narratives inform our work. As sociologist Bruce Podobnik observes, scholars of energy transitions rarely examine the role of labor, civil society, and consumers in the study of energy transitions.[1] But now more than ever, change-oriented groups and individuals deserve credit for challenging powerful energy players.

I have tried to illustrate that energy transition narratives are far more complex than they seem. Yes, mountains are sacred in Maine, New Hampshire, and Vermont, and it is easy to believe that a group of relatively privileged white progressives do not want their mountains defiled by wind turbines. But when citizens with long environmental careers and no personal costs to their property become so disenchanted with wind turbines that they prefer not to see them in *anyone's* backyard, it signals that there is probably more to the story than upper-class angst or selfishly putting the local before the global.

Many of Northern New England's wind opponents express dissatisfaction with the techno-bureaucratic system governing American energy—the way that wind technology is embedded in commercial and regulatory webs. These opponents critique how public-private coalitions prescribe industrial-scale wind turbines as universally essential to energy transition. Without saying so explicitly, many wind opponents are confronting neoliberal economic ideology. They understand that turbines are lucrative investment vehicles and that they may not be as good for the climate as alternatives. Knowing more about the complex political economies activated when consumers turn on their light switches, opponents now demand a more transparent accounting of turbines' carbon reduction potential before simply accepting them.

I wish I could claim that wind resistance has awakened a regional energy consciousness, but that does not seem to be entirely true. As the climate throws more curveballs, the way we think about wind turbines remains alarmingly superficial. Icons allow such autopilot thinking. We react viscerally and instantaneously to the Confederate flag or to a swastika because it takes little cognitive energy to make sense of an uncomplicated and highly recognized symbol. When I was a twenty-year-old college student studying in India, sleeping at the foot of a vertical beam emblazoned with a pink swastika, an ancient Hindu symbol, felt creepy and took some getting used to. Wind resisters have been disabused of the purity symbolized by turbines. Separating the symbolism from the science is something we should all aspire to do in the environmental policy realm. Regardless of wind energy's ecological merits, warm fuzzy associations with wind turbines do little to help create more thoughtful, contemplative energy users.

But we first need energy planning venues that thoughtful energy users can participate in. Right now, the bounds of energy dialogue are narrowly defined by antiquated policies and venues governing electricity. As regulators see it, their job is to smooth the flow of electrons to ensure an adequate and cheap supply, but today this means catering to new groups of private developers, sometimes selling electricity in other states. In energy planning, individuals are called on by regulators to offer only personal feelings about wind turbines. And accordingly, most residents who promote turbines publically speak in favor of the tax benefits or their positive experiences with local developers. Wind opponents must assume equally superficial roles in this process. Despite leveling detailed critiques of industrial wind's climate economics, opponents struggle to launch a strong counternarrative to the hegemonic agenda.[2] Their opposition is still widely understood as a struggle to protect mountaintops. Even the politicians now opposing industrial-scale wind have chosen a rhetorical frame that defends the region's mountains and tourism industry.

I have often resorted to sports analogies and metaphors—not because I like sports, but because I want to make it clear that blaming individuals, equipment, and specific actions in energy debates is myopic. Energy systems are not broken because the benches are stacked with cheaters or because the coaches are inconsistent or inept. Energy systems are broken because the rules were written for different players and different turf. Society has new technological knowledge and social and ecological convictions, and we need new rules. We have never demanded that our legislators and regulators prove our energy sources are better than potential alternatives because spectator participation has never been allowed by the rules. Many voices are being excluded from America's energy transition discourse, intentionally, and in ways institutionalized by laws and tradition.

The difficulty of changing rules is of course our tendency to accommodate and placate those who benefit from the current rule book. Our current renewable energy choices are shaped by the ways that our business
leaders understand and talk about electricity. They broadcast a particular
playbook for transition with shrewd branding and deep penetration in policy circles, and thus we proceed with policies that are neither grounded in
climatological science nor documented with any empirical accountability.
We are conditioned through public energy discourse to see wind turbines
as iconic symbols of energy transition without accounting for enormous
difference in how turbines work in different places. Wind turbines may
indeed have a place in our modern electricity grids, but energy systems cannot be governed by monolithic principles ignorant of geography and climate
science.

Energy Citizenship

The best way to ensure a transition that assigns appropriate weight to the
climate and equity is to allow more voices in energy planning. This equates
to what British social psychologist Patrick Devine-Weight has termed energy
citizenship.[3] Conceiving of an alternative to the passive engagement that
typifies current energy consumers, Devine-Wright advocates for broader rec-
ognition of our energy citizenship as united stakeholders in the ecological
systems under threat by our activities. His concept of citizenship nests
within a large but diffuse global movement that has for several decades
advocated "energy democracy." As scholars Matthew Burke and Jennie
Stephens summarize the concept of energy democracy, it broadly ensures
"public and community control and ownership of the energy sector, while
policies and programs would seek to build capacity for communities to
inclusively and effectively exercise this control for purposes identified by and
accountable to the communities themselves."[4] Under this participatory, civi-
cally engaged model, it suffices to say that climate-focused wind opponents
would have championed markedly different energy strategies from the ones
pursued on their behalf. The following paragraphs contain justifications and
recommendations for building more inclusive energy democracy and more
engaged energy citizens.

Listen, and Make Venues for Listening

To work toward energy citizenship, we must first reform political and regulatory institutions to align with the many goals of energy democracy, which means making these institutions transparent, participatory, and appropriately scaled so that they are inclusive and respectful of diverse perspectives and types of expertise. Energy democracies should not ignore citizen groups who arm themselves with knowledge. More broadly, energy democracies should acknowledge that defense of habitat is an ecological trait, not a selfish human trait. The knowledge gained in defending one's place in the world is no more biased than the knowledge deployed to make money from developing that place. To exclude members of the public with the deepest knowledge of energy systems is to elevate the voices of those with the least knowledge, which is what energy companies want, not what democratic society needs. Democracies must clear the way, through constitutional amendments and regulatory overhaul if necessary, to meaningfully include citizens and local municipalities in energy planning.

Worry about Energy Systems, Not Just Electricity

Even when armed with locally relevant and accurate science, organizing to protest a single iconic energy apparatus has distracted from the unjust and unsustainable impacts of whole energy systems. Debates about wind energy have become venues for exchanging narrow technical expertise—training grounds for attorneys, engineers, and technicians. Rarely do those involved acknowledge the limits of their knowledge or the superficially small parameters of the debate. They traffic in fear, in calculated attempts to have their position regarded as the single valid truth.

I was surprised that many of the well-informed wind opponents with whom I spoke struggled to see themselves as part of a regional energy system, of which electricity was only one part. They could identify the sources of the electrons purchased and distributed by the region's utilities, but only one with whom I spoke, Jim, had an inkling of where the region's gasoline is sourced. It was around the time of the Lac-Mégantic rail disaster, which covered a small town's main street with crude oil from Alberta's tar sands, incinerating dozens of buildings and burning forty-seven people alive. One of the developed world's deadliest train accidents was caused by tanker cars en route to a refinery that distributes gasoline throughout Northern New

England. Jim had always known that oil was refined in New Brunswick, Canada, but not until the disaster did he realize that gas stations all over the region were selling gasoline made from tar sands, transported by rail within two hours of his house.

Science scholar Sheila Jasanoff writes: "Transformative solutions to the world's energy needs will not be achieved without also transforming the ways we look at the problem. For this purpose, it is essential that we take account not only of what science knows but also how science knows it, what it does not know, and how to compensate for our ignorance."[5] In other words, energy solutions will come from breaking down not just the walls between energy insiders and citizens, but also the walls between scientific disciplines, branches of engineering, and other islands of expertise.

Society needs holistic, multisystem-focused energy analysis and planning. Our energy use and our climate crisis are born from multiple systems promoting combustion. At present, our consumption habits rely on innumerable inefficiencies that burn through oil and electrons in transportation, agriculture, architecture, and urban design. For example, we ship our food and the inputs to produce it around the planet, regardless of whether they could be produced locally, and then we waste as much as 40 percent of it. We have organic standards that permit weeding thousands of acres of corn with propane blow torches mounted on tractors. We also use fossil fuel energy in the process of maintaining our physical fitness, driving vehicles to gyms with powered machines to squeeze in workouts made more essential because we spend so much time in vehicles. In Northern New England, electricity constitutes less than 20 percent of CO_2 emissions because, despite a relatively clean electricity portfolio, much of the region must be navigated by car and relies on home heating oil. The region is also known for skiing and dairy farms, both of which consume large amounts of energy in multiple forms. Ski resorts could run their lifts and snow guns on 100 percent renewable electricity but still rely on millions of annual car trips. Dairy farmers are underwritten by petrochemical companies that profit from the fuel in tractors and the fertilizers applied to fields.

We should confront the cumulative ecological costs of these overlapping energy uses with smarter energy systems. A million wind turbines are more ecologically friendly than a billion gasoline engines in cars, but electric motors do not suddenly make cars benign. Electric cars still demand enormous quantities of energy and resources to build. They contribute to road congestion and wasteful idling of other cars, and they release particulates from braking and tire wear that end up in our air and water.[6] Yes, electric cars are long-overdue replacements for the combustion engine, but they are in no way ecologically superior to bicycles or electric trains and buses. Of course, there are cascading personal and social benefits from transit that is more human-powered and collective.

Simply put, energy policy goes well beyond electricity policy. Finnish scholars Atte Harjanne and Janne Korhonen suggest that when society puts so many eggs in the renewable electricity basket, needed attention is distracted from climate-friendly transformations. Harjanne and Korhonen suggest that "one beneficial way to reframe the discussion could be to shift the focus from conceptualizing energy systems through energy production methods—which also lets other sectors off the hook, so to speak—to conceptualizing sustainable, low-carbon societies and what they might look like in reality."[7]

Until Energy Is Clean and Fair, Pay Its True Cost

Local energy democracy cannot remedy an energy system designed to sustain gluttonous consumption because climate change affects all backyards, and participation locally does not automatically secure just and sustainable outcomes globally. Despite the speed and ease of global travel and communication, we are still motivated most strongly by threats that are local, immediate, and personal. Worse, the way that our debates unfold, in states and nation states, distracts us from how our energy decisions impact people across borders. Political boundaries do not stop pollutants or negate our moral responsibility as energy consumers for those living far away.

Conceiving of an energy transition in which the costs and benefits are distributed fairly requires owning up to rich countries' disproportionate energy use and its many consequences. The nations with the largest ecological footprints are not just burning through their own stores; they are also consuming finite resources abroad. Whether involving oil for electricity or transport, sand for concrete, water exported in bottles or retained in crops, or components for wind turbines and solar panels, the material flows that enable consumer society send ripples of great consequence to the farthest corners of the planet. We should be concerned about the faraway toxic mines and miners needed to make renewables and batteries, and about what happens to renewables and batteries when they reach the end of their useful lives. If we truly care about justice, we should not let the fanfare surrounding cost-competitive renewables obscure Western society's sense of entitlement to the world's natural resources.

Entwined with the argument that electricity must be cheap is the ideology that our consumption of it need not decrease; but as long as the electricity grid accepts electrons from fossil fuels or otherwise spurs exploitation, that ideology is antithetical to climate change mitigation or justice. Of secondary concern are the unsustainable levels of consumption that low-cost energy facilitates from the things we plug in or buy with the money cheap energy saves us.[8] Emissions from our electricity have never been the only negative externality associated with Western lifestyles. The cheap energy imperative generates layer upon layer of wasteful ecological relationships and is a shackle on American climate policy. To be clear, the issue is not that expensive energy is good for society as a rule; the issue is that right now cheap energy has severe and far-reaching consequences.

Holding ourselves to the highest possible social and ecological standards may entail overhauling energy infrastructure, and we may find that the most effective means to sustainably restructure our energy systems requires consuming less electricity. However, it would be playing into the hands of energy companies to believe that consuming less electricity means sacrificing creature comforts. As research from Australia shows, linking transition to economic sacrifice and deprivation is a rhetorical tactic called "discursive misdirection" used by energy companies to stoke fear in consumers.[9]

An alternative scenario is rarely discussed; we could be paying more for our electricity, using less of it, and sacrificing nothing in terms of our daily comfort by fully reimagining energy systems, not just swapping out energy sources. The current system orients us toward a flow of electrons rather than the things in life we actually want. As several engineers and design scholars have noted, the simple things that we use electrons to accomplish are the actions that reward us and make our lives better: washing our clothes and dishes, charging our phones, streaming movies and music, powering hospitals.[10] No consumer derives a sense of fulfillment or satisfaction from the number or cost of electrons coursing through their circuits. Those sentiments belong entirely to people selling electricity.

Of course, a just energy transition should aid the most vulnerable victims of climate change and the poor who struggle to afford their basic daily needs. If the consequence of rapidly transitioning from fossil fuels is paying more for electricity, we can assist low-income households while also aggressively pursuing energy conservation. For example, we can implement progressive energy taxes tied to emissions, with the highest burden on the most gluttonous uses and users. Many policies could be restructured to make efficient transportation, heating, and weatherproofing for poor households more attainable. Additionally, it seems both logical and justified to start a fund for achieving these goals by holding energy companies accountable for their long ecological con, following the precedent set with tobacco settlements.

How can local energy planning remedy global injustice? Energy proposals could be evaluated for human impacts universally. At the very least, proposals should require a transparent accounting of the local and global distribution of carbon costs and benefits, mandate fair and safe sourcing and recycling, and apportion a far more significant share of energy profits to be used toward stringent efficiency goals.

The causes of inequality and social justice can be placed on equal footing in energy planning. Environmental sociologist David Hess uses the term "access pathways" to describe points of convergence where social and environmental initiatives can collaborate toward shared causes. Hess offers the examples of sustainability and hunger prevention, and affordable housing and weatherization, among others, contending that "green access may be considered an unaffordable luxury, but thinking through what it can involve could help ignite powerful synergies between the social justice and environmental sustainability agendas."[11] Legislators in cold rural places, if they wish to democratize energy and energy planning, should consider clearing the "access pathways" of transit, home heating and efficiency, to help grow—not shrink—the number of households actively participating in efforts to address climate change.

Think Small

When sorting through points of convergence and disagreement in the wind debate, it is clear that many in Northern New England are passionate about community scale. Yes, the wind is renewable, but a four-hundred-foot ridgeline wind turbine has about as much in common with a one-hundred-foot turbine as a fifty-cow dairy farm has in common with a ten-thousand-cow dairy farm. While scale is good for accumulating money in one place, doubt lingers as to whether scale can achieve sustainability. Industrial wind and the massive grid to which it feeds electrons exemplify Beck's observation that scale brings risk. Beck argued that the scale of modern systems exposes them to risks becoming steadily more frequent. Vast networks of overhead electricity lines leave power systems vulnerable to weather and human sabotage, just as a centralized supply chain introduces risk to food systems.

Scale also entails severe climate risk. Moving electrons farther creates more opportunities for infrastructure to fail. Moving electrons farther also creates the need for ecologically costly upkeep. A 2019 *BBC News* report decried the electricity grid's reliance on sulfur hexafluoride (SF_6), a greenhouse gas 23,500 times more heat-retaining than CO_2. SF_6 keeps electrical components from suffering catastrophic failure, but inadvertent leaks in the United Kingdom in 2017 added the equivalent of 1.3 million cars of CO_2 to the atmosphere.[12] As more grids accept diverse sources of electricity, they demand more components that risk failure, increasing the need for more gases like SF_6. Large grids also result in electricity losses when transmitting electrons long distances, requiring more electricity generation than if all electrons were used locally. Yet, with the exception of Vermont's Green Mountain Power, few others embrace a future of microgrids in which energy generation is distributed to meet local needs, minimizing the need for regional interconnection.

Green Mountain Power's innovative and ambitious vision of distributed microgrids shows that "alternative pathways" can emerge from creative thinkers in the private sectors. But ordinary citizens, communities, and neighborhoods can also troubleshoot local energy puzzles.

Rather than wait for policy makers to more aggressively target carbon emissions, wait for more accessible energy planning venues, or wait for revolutionary technological breakthroughs, individuals and community groups who desire more active engagement in energy issues can work at the margins of centralized energy systems to reform them from the outside. Geels contrasts the tendency to rely on green innovations born from industrial-scale research and development to models of small, local work creating, testing, and tinkering in the field to solve energy problems.[13] Geels likens this scenario to the Bible's David, slowly weakening, eroding, and destabilizing Goliath, to "enhance the chances of green Davids."

To get started, there are established models around the world worth replicating, refining, and building on. They include bulk buying clubs for residential energy apparatuses like solar panels and batteries, community and cooperatively owned microgrid projects, and energy clubs and volunteer organizations to assist with retrofitting. Inspiring examples emerged in Puerto Rico after Hurricane Maria crippled the island's electric grid. In Northern New England, community energy committees and clubs are an example of potential pathways.[14] Without trying, ventures like these organized from the ground up have been able to claim space from hegemonic energy interests because they are unencumbered by the mindset that electricity must flow downward to the people from above.[15] And as Hess suggests, forging community capacity around energy can be the catalyst for spinning off bigger structural changes to tackle things like education, food systems, and the brain drain.

Final Thoughts

Citizens should move fast. Geels warns that although "innovations have more sustainability promise, such a shift in the mode of innovation is likely to be resisted by incumbent regime actors."[16] Evidence of such resistance is easy to find. Energy policy makers are also aware of unfolding changes to scale, but not in a way that promotes democracy, shared ownership, or energy justice. Commissioner of the Federal Energy Regulatory Commission, Cheryl LaFleur, told attendees at a microgrids conference in 2018 that America's outdated electricity policies are delaying distributed energy generation and community-shared heating and cooling.[17] Sounds hopeful, right? The story then describes that private investors are poised to devote loads of capital to such energy projects but are reluctant to do so with no clear policy to protect their investments. This shows that advocating for microgrids is not necessarily an argument to dismantle an energy system that gives perpetual home field advantage to large companies. It is sometimes just an argument to let a few new players on to the field.

Rather than worrying about venture capitalists' lost opportunities, renewable energy debates underscore the need for energy policy that empowers individual ratepayers to actively and meaningfully participate in policy redesign. Simply opening up new business concepts to investment will do very little to close the knowledge gap that keeps consumers captive to structures that make us energy addicts while rewarding distant shareholders. New investments will further incentivize selling electricity by volume and thus weaken the mechanism through which consumers would logically exert demand for more efficient, and thus more just, homes and gadgets.

We can see why energy is a problem desperate for democratizing when we think about all the low-hanging fruit we are leaving on the vine in terms of efficiency. Consider a December 2018 article in the *Wall Street Journal*, heralding the momentous efficiency potential that building technologies can already achieve. Titled "Why Your Next Home Might Not Need Any Energy at All," the article catalogs advancements in design, materials, and construction technique that dramatically reduce buildings' thirst for energy.[18] Erasing heating, cooling, and lighting expenses would make small residential solar arrays even more viable and challenge many assumptions about how much energy society actually needs.

A March 2019 *New York Times* op-ed tempers this good news with a snapshot of the democratic obstacle course standing in the way of design progress.[19] The apparatus for enforcing energy efficiency in buildings, the International Energy Conservation Code, is amended periodically when cities and towns vote on changes, but this recurring ballot often falls to the bottom of municipalities' long list of priorities. Well-funded homebuilders' groups, on the other hand, lobby hard against code improvements that require more time and money on their part. To summarize, one of the most effective ways to harness energy consumption without sacrificing comfort would be to utilize already available materials and know-how in future construction, and yet even when confronting climate change's catastrophic consequences, our society lacks the organizational capacity to pick such low-hanging fruit. If concerned citizens wielded even half the energy influence held by companies, a more holistic energy policy might be attainable.

If done right, a large-scale policy such as a "Green New Deal" could spark the energy citizen revolution, and with less pain than a sudden energy shortage would, but a "Green New Deal" would have to pivot dramatically from past efforts that threw money at slow-transition interests. If instead of subsidizing industrial scale renewables the federal government instituted a nationwide carbon tax to fund locally calibrated programs that maximize carbon reduction per dollar, what would the Northern New England states do? I suspect that grid-scale onshore wind turbines would not be sited in the same places or quantities. Industrial wind turbines would probably still cluster in windy places like Texas and the Southwest, where hot summer weather requires electricity for air-conditioning and constitutes a large portion of those regions' carbon emissions. But Northern New England would use its funds to winterize its buildings, maximize passive design, and put fewer gasoline-powered cars on the road. These uses would undoubtedly put more people in jobs by benefiting an array of industries, all of which are relatively small, independent, and decentralized. Construction trades; heating, ventilation, and air-conditioning (HVAC) contractors; and transit services would benefit, as opposed to international utilities and their share-holders. Perhaps energy clubs and energy committees would spin off small businesses, creating more and greener local jobs than grid-scale wind. Northern New England's renewable electricity would probably originate from publicly financed and community-owned solar and low-impact microhydro projects, perhaps tied to hydrogen production if not natural storage. The need for renewables would be moderated by aggressive initiatives to curb overall energy use, which are all more cost-effective than large-scale wind, though not as grandiose and totemic. Realistically, under any scenario the region will continue to buy hydropower from Quebec, at least until localized microgrids proliferate—but rather than thinking of Hydro Québec's indigenous victims as an issue from which we should distance ourselves through energy independence, the region should engage and collaborate with them to be levers of justice in their struggle.

The world would be well served by an energy future in which a more level playing field for electrons co-evolves with a more level playing field for people. Modern renewables have granted rural communities a new level of political and economic relevance at a time when rural livelihoods face existential threats and rural culture is fracturing. Voters, leaders, and rate-payers could leverage that relevance to promote an energy transition that is not just ecologically restorative, but also socially and politically transformative. As we learned from rural electrification in the 1930s, rural places make excellent proving grounds for experiments of democratization.

Epilogue

●●●●●●●●●●●●●●●●●●●●●

In the six years that I have studied renewable energy in Northern New England, much has changed and much has stayed the same. All three states have seen major ridgeline wind projects come online. And all three states, despite different political climates, have seen decreasing openness to new ridgeline wind projects. Senator Rodgers's wind turbine moratorium did not gain sufficient traction to become law in Vermont, but development has stopped for now anyway. In 2020, Maine is the only of the three states experiencing continued wind turbine construction. All three states have seen projects fail at multiple steps along the path to approval—some because of scenic protections imposed by regulators, and some because of utilities' ambivalence about additional wind-generated electrons. Among national renewable energy developers, the Northeast now has a reputation for its discord, which makes outside investors interested in pursuing renewable energy projects elsewhere, where wind will displace coal and therefore reduce more CO_2 emissions.

The tenor of the region's renewable energy debate remains rancorous. Despite overhauling its controversial SPEED program and tightening restrictions on turbine noise, Vermont leaders remain under intense scrutiny from community groups. With liberal rules for interstate REC trading, Vermont's 2015 reform legislation did little to address the perceived injustices associated with rural communities playing host to the land-intensive energy needs of cities and suburbs. Municipalities had sought the authority to keep new industrial-scale wind energy projects at bay but instead got Act 174. Passed in 2016, the legislation mandates that towns designate parcels appropriate for energy-generating infrastructure, which attracted several contentious large solar projects by out-of-state speculators capitalizing upon the law's ambiguity around scale, ownership, and the distribution of benefits. Two municipalities were sued by developers, angry that their preferred sites for solar installations were not designated as such. Because renewable energy currently provides relatively little electricity, and because so much of it goes out of state, many believe that Act 174 should not be confused with "local energy planning."

Just to the west, in the state of New York, the same issues persist, just under very different geopolitical conditions; highly urban and remote rural populations are united in the same state. New York City's elected officials are showing new resolve to eradicate fossil fuel combustion from the urban communities shouldering an unfair pollution burden, and the state legislature is designing aggressive renewables policies to buildout rural wind and solar capacity. These policies are of course eliciting backlash from the rural host communities being tapped to host renewable infrastructure but being promised fewer tax benefits for doing so thanks to generous incentives funneled to developers. Vermonters, no matter their distaste for their renewables policies, are probably grateful they do not answer to politicians in Albany.

In 2020, Vermont lawmakers finally targeted climate change explicitly with legislation called the Global Warming Solutions Act. The law outlines emissions targets and a transition from fossil fuels by 2030. Another provision places an arbitrary cap on the proportion of hydropower purchased from Quebec; this provision suggests that lawmakers want desperately to sustain existing solar and wind jobs, which is proving to be a struggle. Vermont news headlines have tracked declining solar installations and solar jobs, attributing it to continued net-metering rollbacks, Trump's tariffs on solar panels manufactured in China, and now COVID-19. The new law, passed in spite of the governor's veto, allows parties to sue the state for failing to meet its targets. Utilities have warned that the law will cause electricity rates to soar. Environmentalists have decried it because the Climate Council it authorizes is dominated by business-as-usual energy interests.

In 2020, a major rollback of the federal regulations that first gave rise to renewable energy threatens solar and wind more widely and more acutely.

Amendments to the Public Utility Regulatory Policies Act (PURPA) will give states more leeway to regulate and allow utilities to pay lower rates when putting renewable electrons on the grid. Utilities claim that the amendments will rope in sweetheart deals that left ratepayers on the hook for paying renewable developers' enormous profits. Renewable developers see the amendments as a boon to fossil fuel interests, further crippling our energy transition. The new rules will allow utilities more latitude, which in a progressive regulated market like Vermont will not be nearly as detrimental for the climate as in other states. What is clear is that these rule changes restore primacy to utilities in setting prices, and while lowering costs was the battle cry that won utilities favor with federal regulators, low costs are great for utilities, which sell by volume, and horrible for the planet so long as fossil fuels exist on the grid.

With more space and more sparsely settled terrain, Maine's wind development has advanced with less controversy following the 2018 elections. Two projects with a combined total of twenty-six turbines proposed before Governor LePage's moratorium have advanced to construction phase, and a few additional proposals wait in the wings. The state's proposed floating turbines off the coast of Monhegan Island stalled amid concerns that too many costs would be shouldered by ratepayers, but new developers anticipate completion by 2023. In addition, Massachusetts's calls for carbon-free energy have spawned several already mapped-out ridgeline wind projects in Maine to advance, including expansions of existing installations with newer larger turbines, and installations at new sites that might tap into a proposed undersea cable delivering power directly to Boston. Despite the prominent role that local energy coalitions played in priming the cultivation of a state wind industry, the developers advancing plans in Maine are all from out of state, as far away as Virginia and Texas.

Other energy issues in Maine continue to make headlines and stir controversy. Central Maine Power, owned by Spanish conglomerate Iberdrola, earned the lowest customer ranking of any electric utility in the country because of erratic rates and billing errors. The state's second monopoly utility, Emera Maine, was sold in 2020 to ENMAX, a municipally owned electric utility in Calgary, Alberta. ENMAX borrowed heavily to finance the sale, raising fears about skyrocketing rates coming down the pike. This tumult has culminated in contentious political maneuvering. In her 2020 State of the State address, Governor Janet Mills made overtures suggesting she hoped to re-evaluate the regulatory landscape in which Maine's energy decisions are made. While the governor's ambitions have so far remained vague, Seth Berry, a fellow Democrat in the legislature, funded a study to determine whether it would benefit Maine residents to wrest control of its two investor-owned utilities from their foreign owners. It seems reasonable that if a public entity is to own Maine's electricity interests, it might as well be the state of Maine and not the city of Calgary.

Of continuing importance to Maine's energy dialogue are proposed transmission projects to carry hydroelectricity from Quebec to Massachusetts. New Hampshire's defeat of the Northern Pass immediately put Maine in the crosshairs because Central Maine Power had eagerly charted a route through northern Maine well in advance. Maine residents have responded in force, with signs, bumper stickers, op-eds, and two drives for a statewide referendum. The first referendum's wording was ruled unconstitutional by the Maine Supreme Court in August 2020. A second referendum was gaining signatures in early 2021, even as crews readied construction equipment deep in the Maine wilderness. The "New England Clean Energy Connect" seems likely to proceed, though legal challenges will probably delay it. What is certain is that it has stoked enormous publicity and debate. The project elevated to public prominence the scientific uncertainty surrounding large-scale hydropower's ecological footprint. Also, Hydro Québec's undisclosed role in promoting the project in the media resulted in sanctions, and a judge later forced citizens groups opposing the project to disclose their donors as well. One opposition group disclosed over $4 million in contributions from fossil fuel power plant owners, prompting a local television station to run the headline: "Big money from away flows in rival Maine power line campaigns." CMP has itself spent $13 million on its ad campaign for the project. While the company has promised hundreds of millions of dollars in benefits to ratepayers and communities, opponents find these promises paltry relative to the profits CMP will reap for foreign shareholders. It is projected to net $5 million per month from the corridor project once completed.[1]

Unlike its neighboring states, New Hampshire's carbon reduction needs are less pressing. The state's nuclear facility is licensed to continue operating until 2030, and it has applied to extend its license to 2050. New Hampshire is also home to several gas plants and hosts Northern New England's last two coal plants, used primarily for peak loads in the summer and winter. That New Hampshire pursued wind installations at all is testament to the industry's appeal to investors at the height of the stimulus building boom.

Nonetheless, recent energy headlines in New Hampshire portray ongoing tension between big and small energy interests. The state's Republican governor has vetoed legislation to open net-metering for solar installations to municipalities and businesses, which reporters speculate relates to lobbying from the fossil fuel industry. In typical neoliberal fashion, Governor Chris Sununu is particularly resistant to allowing municipalities to net-meter, but New Hampshire utilities and the state's solar industry had advanced the vetoed legislation through compromise. As terms of the compromise, regulators commissioned the study of net-metering's impacts on utility expenses, which will provide much needed clarity to the debate over whether solar saves or costs money for utilities and ratepayers. (Like in neighboring states, New Hampshire authorities strongly reject the 2020 report finding billions of savings attributable to solar across New England, and still await the results of their own study.) Several New Hampshire municipalities, meanwhile, are organizing to wrest control of electricity pricing and sourcing from utilities, by essentially becoming group buyers on the deregulated market. The proposed legislation allowing for "The Community Power Coalition" is of course controversial. Another bill under consideration in New Hampshire's legislature in 2021 seeks to lower the costs and expand due process for wind turbine neighbors who file grievances when wind projects violate the conditions of their permits. The bill would lower the filing fee from $10,500 to $250, and redirect adjudication from a single administrator to the entire Site Evaluation Committee.

Despite their different energy markets, political traditions, and parties in power, the three Northern New England states appear surprisingly uniform in their social responses to ridgeline wind and the eventual retrenchment by legislators. Public hearings on energy issues continue to last late into the evening. Groups continue to meet in person and in online forums. They share news and commentary on social media pages. The red disco lights atop turbines still blink frenetically at night. It turns out the technology to activate them only in the presence of airplanes, promised for ten years, is incompatible with the terrain.

Acknowledgments

This book would not be possible without those who generously agreed to talk with me when I knew very little about energy, answering my questions without hesitation. Your insights form the core of my work, and I hope it does them justice. Similarly, I could not have achieved the same depth of insight on regional energy without the dedicated individuals working throughout New England to raise awareness about energy issues. Your efforts to share news, maintain social media accounts, and video record public meetings was tremendously useful for me. In particular, my inquiry was made far richer by the reporting and forum-hosting by the *VT Digger*. The work you do is outstanding and Vermont is a better place because of it.

I am immensely grateful for the advice and encouragement from my manuscript's two reviewers, who saw promise in material that few others had looked at. Your perspectives were unique, valuable, and key to many improvements. I am also appreciative of my students in my Pipelines course who read and edited an unpolished draft in spring 2019. I am grateful for financial support from the faculty development funds of Bowdoin College and Kenyon College, from the U.S.-Norway Fulbright Foundation, and from the Center for Resource Solutions, which granted me access to its conference.

I envisioned writing a book long before I ever knew what social research was, and so I have many individuals to thank for preparing, encouraging, and nourishing me over many years.

From Kenyon, I am thankful for my outstanding colleagues in the Department of Sociology and across: in particular, Austin Johnson, Gillian Gualtieri, Jodi Kovach, Katie Mauck, Marla Kohlman, Martha Gregory, Matt Suazo, and Rakia Faber.

From Bowdoin, I thank Kelly Fayard, for keeping me on course with levity and emotional support, and for reading and offering feedback long before anyone else. I am grateful to Ingrid Nelson, my friend and sociological conspirator for more than twenty years, who has helped me as a scholar immeasurably. I thank Monica Brannon for sharing her wisdom on the study of technology and power, and for wise counsel on all professional matters.

From the University of Wisconsin–Madison, I am grateful for the continued advising, support, and comradery of Katherine Curtis; for Jane Collins, who encouraged my work and whose course transformed my understanding of the social sciences; and for Mary Layoun, whom I admire tremendously and whose decision to hire me in 2003 kept me from dropping out. My friends, classmates, and housemates from Madison are an enduring source of grounding and inspiration. They include Kat Becker, Nicole Breazeale, Lena Etuk, Jenn Huck, Brent Kaup, Sarah Lloyd, Amy Quark, Becky Schewe, Tony Schultz, Adam Slez, Jude Toche, Paul Van Auken, and Richelle Winkler. I could never forget Bryan Miller, my dear friend and inspiration for applying to graduate school, who kept me alive and laughing on the Mongolian Steppes. I am also grateful for Johan Fredrik Rye and his colleagues at Ruralis and NTNU for many years of support and collaboration.

From my undergraduate days, I have deep appreciation for my Bowdoin professors Jill Pearlman and Joe Bandy. Your courses were formative, and since finding my way to this profession, I have tried to replicate them. I am also forever grateful to my IHP family, with whom I spent eight incredible months thinking about big problems and small solutions.

I owe an enormous debt of gratitude to those who encouraged me, taught me to write, and to think critically and compassionately in the public schools of Weathersfield and Windsor, Vermont. Your classrooms opened opportunities too numerous to count. Thank you especially to Marilyn Yushak, Joan Holzwarth, Ginger Wimberg, Muriel Roland, Peter Berger, Kathy Taylor, Carol Morgan, Beth Dutton, and Bill Stanard. I am also grateful for the staff of the Keene State College Upward Bound Program for helping me set my sights high. Know that your work was profoundly influential

I thank my father, Philip, my champion and unwavering supporter through my best and worst. My sociological imagination probably comes from you. I am also grateful for my mother, Cheryl, who in memory has been my compass, and for my siblings, nieces and nephews; my Aunt Margaret; and my stepmother, Teresa.

I am fortunate and grateful to have several surrogate families whom I love as my own: the Babbs, the Casey-Nelsons, the Etuk-Kruzics, the Holmans, the Karlsaune-Grovassbakks, the Parkers, the Tobey-Pyburns, and the Van Vleck-Reillys.

Notes

Chapter 1   Introduction

1 Beder, *Power Play*; Hughes, *Networks of Power*.
2 Beckley, "Energy and the Rural Sociological Imagination."
3 Verbong and Geels, "The Ongoing Energy Transition"; Geels, "Regime Resistance against Low-Carbon Transitions"; Hess, "The Green Transition, Neoliberalism, and the Technosciences."
4 Bell, "Environmental Injustice and the Pursuit of a Post-Carbon World"; Goldstein, *Planetary Improvement*; Mulvaney, *Solar Power*; Sovacool et al., "Valuing the Manufacturing Externalities of Wind Energy"; Zehner, "Unintended Consequences."
5 York and Bell, "Energy Transitions or Additions?"
6 Harjanne and Korhonen, "Abandoning the Concept of Renewable Energy."
7 Morris and Jungjohann, *Energy Democracy*; Devine-Wright, "Energy Citizenship"; Burke and Stephens, "Political Power and Renewable Energy Futures."
8 Toke, *Ecological Modernisation and Renewable Energy*; Vasi, *Winds of Change*.
9 Interviews were conducted between May 2013 and August 2016, throughout Maine, New Hampshire, and Vermont. All interviews were audio recorded and transcribed. Quotations attributed to interviewed respondents herein are presented verbatim. Interviewed respondents have been assigned pseudonyms.
10 Quotations presented herein that appeared online in public comments are presented verbatim. Quotations are not attributed to legal names or pseudonyms unless the commenter disclosed their identity as a public official.

Chapter 2   Windy Ridgelines, Social Fault Lines

1 Polling estimates in Vermont placed voter support for wind energy near 70 percent (Castleton University Polling Institute, "Complete Poll Results"). In

Maine, support for wind was estimated to be higher than 80 percent (Turkel, "Mainers Full of Gusto for Wind Power").

2 It is noteworthy, however, that none of the region's small turbine manufacturers produced equipment appropriate for grid-scale installation locally, severely limiting the local economic impact.

3 In 2010, Vermont's legislature transferred appeals of PUC decisions from environmental court back to the PUC, with second appeals heard by the state supreme court.

4 Stein, "Shumlin Administration Forms Commission to Assess Siting Process."

5 Anair and Mahmassani, *Clean Energy Momentum*.

6 Guy and Shove, *The Sociology of Energy, Buildings and the Environment*.

7 Devine-Wright, "Beyond NIMBYism."

8 Devine-Wright, "Beyond NIMBYism."

9 Van der Horst, "NIMBY or Not?"

10 Toke, Breukers, and Wolsink, "Wind Power Deployment Outcomes"; Farrell, *Democratizing the Electricity System*; Morris and Jungjohann, *Energy Democracy*.

11 These "infra-sounds" are the subject of great debate, but, for clinical purposes, many in the medical community have acknowledged a suite of symptoms broadly associated with living near wind turbines as "wind turbine syndrome."

12 "Wind Turbines: Do Property Values Fall?"

13 Predictably, lowered property values appear to impact only close neighbors, and therefore studies covering large areas may blunt the effect. Also, wind developers in Northern New England have made buyouts and buyout offers to stem litigation brought by wind neighbors, which may make losses to home values more difficult to track.

14 Angwin, *Shorting the Grid*.

15 Wilson and Stephens, "Wind Deployment in the United States"; Toke, Breukers, and Wolsink, "Wind Power Deployment Outcomes"; Krogh, "Industrial Wind Turbine Development and Loss of Social Justice?"; Wexler, "A Sociological Framing of the NIMBY (Not-In-My-Backyard) Syndrome"; Jacquet, "Landowner Attitudes toward Natural Gas and Wind Farm Development."

16 Beder, *Power Play*.

17 Benford and Snow, "Framing Processes and Social Movements."

18 McClymont and O'Hare, "'We're Not NIMBYs!'"

19 Kallman and Frickel, "Power to the People."

20 Devine-Wright, "Public Engagement with Large-Scale Renewable Energy Technologies."

21 Wolsink, "Entanglement of Interests and Motives"; Devine-Wright, "Beyond NIMBYism"; Freudenburg and Pastor, "NIMBYs and LULUs"; Hager and Haddad, "A New Look at NIMBY."

22 Moore, "Defining 'Successful' Environmental Dispute Resolution"; Smith and McDonough, "Beyond Public Participation"; Lachapelle and McCool, "Exploring the Concept of 'Ownership' in Natural Resource Planning."

23 Parkins et al., "Can Distrust Enhance Public Engagement?"

24 Devine-Wright, "From Backyards to Places."

25 Schreurs and Ohlhorst, "NIMBY and YIMBY," 79.

26 Pepermans and Loots, "Wind Farm Struggles in Flanders Fields."

27 Freudenburg and Pastor, "NIMBYs and LULUs."

28 Wüstenhagen, Wolsink, and Burer, "Social Acceptance of Renewable Energy Innovation."
29 Noble, "Wind Won't Be Our Savior."

Chapter 3 For the Love of Mountains

1 Goldman, "James Taylor's Progressive Vision."
2 Brown, *Inventing New England*.
3 Cronon, *Changes in the Land*.
4 Judd, *Common Lands, Common People*.
5 Vermont Commission on Country Life, "Rural Vermont."
6 Mills, "The Unheralded Champion of Maine's 'Vacationland' License Plate Motto."
7 Sherman, Sessions, and Potash, *Freedom and Unity*.
8 Brown, *Inventing New England*.
9 Morrison and Wheeler, *Rural Renaissance in America?*; Beale, "A Further Look at Nonmetropolitan Population Growth since 1970."
10 Brown, *Inventing New England*; Harrison, *The View from Vermont*.
11 Brown, *Back to the Land*; Jacob, *New Pioneers*.
12 Golding, "Rural Gentrification in the United States 1970–2000."
13 Kuentzel and Ramaswamy, "Tourism and Amenity Migration"; Kacprzynski and Graefe, *Proceedings of the 1990 Northeastern Recreation Research Symposium*.
14 Trust for Public Land, *New Hampshire's Return on Investment in Land Conservation*.
15 In tandem with the wave of new federal protections for air and water quality, several state-level policies enacted in the 1970s and 1980s redoubled Northern New England's efforts to reduce pollution and promote conservation and environmentally responsible development. Maine and Vermont implemented bottle bills to clean highways and incentivize recycling. The two states also restricted development through comprehensive oversight of proposed developments. Both states also instituted bans on billboards, distinguishing Maine and Vermont from all but two other U.S. states, Alaska and Hawaii. These accomplishments have remained in effect through leadership by both Democratic and Republican governors and legislatures. Conserving forests and forest economies is also a consistent priority. While each state earmarks public funds for conservation, a sizable portion of the region's land conservation relies on private philanthropy working through nonprofit land trusts. According to data released in the 2015 National Census of Land Trusts, the three states rank in the top five U.S. states for number of land trusts per capita and lead the nation in the percentage of the states' total land area conserved through land trusts. Just as with renewable energy, Vermont leads the way in terms of landscape preservation and conservation. Vermont was among the first states to implement statewide environmental permitting standards in the early 1970s, and in 2015 Vermont began enforcing the strictest recycling laws in the nation, which forbid the disposal of biodegradable materials in landfills.
16 Albers, *Hands on the Land*; Morse et al., "Performing a New England Landscape."
17 Dillon, "Activists Press for Close Review of Vermont Utilities' Canadian Ownership."

18 Sabourin, "In Quebec, Canada's Newest Hydroelectric Dams Nearly Ready."
19 Devine-Wright, "Think Global, Act Local?"
20 Phadke, "Public Deliberation and the Geographies of Wind Justice."
21 Woods, "Deconstructing Rural Protest."
22 Halfacree, "Contrasting Roles for Post Productivist Countryside."
23 Marcouiller, Kim, and Deller, "Natural Amenities, Tourism and Income Distribution."
24 "Big Wind, Small State."
25 Pepermans and Loots, "Wind Farm Struggles in Flanders Fields"; Batel and Devine-Wright, "A Critical and Empirical Analysis of the National-Local 'Gap.'"
26 Burke and Stephens, "Political Power and Renewable Energy Futures."
27 Farrell, *Democratizing the Electricity System*.
28 Cronon, *Changes in the Land*.

Chapter 4   But What If . . . ? Wind and the Discourse of Risk

1 Firetrace, *In the Line of Fire*.
2 Beck, "On the Way to the Industrial Risk-Society?"
3 Beck, "On the Way to the Industrial Risk-Society?"; "Risk Society and the Provident State"; and "Climate for Change."
4 Museum of Science and Industry, "Einstein May Outrank Britney Spears."
5 Shultz, "The Renewable Energy Rebels."
6 Herrick, "Green Businesses Press for Carbon Tax."
7 Miller and Keith, "Climatic Impacts of Wind Power."
8 Li et al., "Climate Model Shows Large-Scale Wind and Solar Farms in the Sahara."
9 Way, "Duke Energy Application Points Finger at Solar."
10 Miller and Keith, "Observation-Based Solar and Wind Power Capacity Factors and Power Densities."
11 Luce, "Issues with Wind in Our Region?"
12 Holland and Luce, "The Cost-Effectiveness of 21 Vestas V112 Wind Turbines."
13 Dillon, "Vermont's Energy Credits Questioned."
14 Parker, "Public Service Commissioner."
15 Sovacool et al., "Valuing the Manufacturing Externalities of Wind Energy."
16 Shemkus, "Maine Transmission Line Entangled by Conflicting Claims on Emissions."
17 Hager, "Commentary: Hydro-Québec Offers Misleading Claims."
18 Beck, "Climate for Change."

Chapter 5   Following Power Lines

1 Vermont Department of Public Service, "2020 Annual Energy Report."
2 Righter, *Wind Energy in America*.
3 Hess, "A Political Economy of Sustainability," 244.
4 Harvey, *A Brief History of Neoliberalism*.
5 Mirowski, "Neoliberalism."
6 Harvey, *A Brief History of Neoliberalism*.

7 Dickinson, "How Big Oil and Big Soda Kept the Plastics Crisis a Secret for Decades."

8 De Wit and Bigaud, "No Plastic in Nature."

9 Trucost, *Plastics and Sustainability.*

10 Regulated electricity markets like Vermont's, where central authority has been used to expand renewable energy, help to illustrate why neoliberal electricity policy can be hard to generalize in terms of environmental responsibility. The political appointees charged with overseeing Vermont's electricity markets utilize the state's lingering electricity monopoly to undertake and underwrite large and expensive development projects. Vermont's Sustainably Priced Energy Enterprise Development (SPEED) program, which ended in 2015, incentivized wind turbine development by guaranteeing renewable energy developers long-term electricity rates for the output of their new projects, a popular neoliberal policy archetype.

11 Sovacool, Gilbert, and Thomson, "Innovations in Energy and Climate Policy."

12 Hess, "Electricity Transformed," 1075.

13 Hess, "Electricity Transformed," 1076.

14 Angwin, *Shorting the Grid.*

15 Angwin, *Shorting the Grid.*

16 State tax incentives for wind developers exist in a handful of U.S. states but not in Northern New England, although local municipalities have been cleared to offer the incentives if desired. In eastern Maine, one of the poorest parts of the American Northeast, a wind developer is in negotiations with municipalities to incentivize a thirty-turbine project using a special program called tax increment financing (TIF). TIF status is reserved primarily for economically challenged places. It incentivizes development by promising developers future tax reductions. If approved, the developer in eastern Maine would keep 70 percent of the property tax levied on its wind installation, distributed over thirty years.

17 O'Shaughnessy et al., *Status and Trends in the U.S. Voluntary Green Power Market (2016 Data).*

18 Post et al., "Petition to Investigate Deceptive Trade Practices of Green Mountain Power Company."

19 Jones, James, and Huebner, "Do You Know Who Owns Your Solar Energy?"

20 Dillon, "Transmission Grid Bottlenecks in Northeast Kingdom Stall Solar Development."

21 MacNeil and Paterson, "Neoliberal Climate Policy."

22 Varoufakis, *Talking to My Daughter about the Economy,* 132–133.

23 Knight and Hall, *Wholesale Cost Savings of Distributed Solar in New England;* Knight et al., "Hourly Price Impacts of New England Solar."

24 Anderson, "Utilities, Koch Groups Back Sununu's Veto of Net Metering Bill in NH."

25 Aklin and Urpelainen, *Renewables.*

26 Bakke, *The Grid.*

27 Angwin, *Shorting the Grid.*

28 Kantamneni et al., "Emerging Economic Viability of Grid Defection in a Northern Climate."

29 Hess, "A Political Economy of Sustainability," 251.

30 A trade group in Vermont, "Businesses for Social Responsibility," dates back to 1990, and in 1995 its membership conceived of a "Sustainable Jobs Coalition,"

which earned legislative support and state funding. In chartering the Sustainable Jobs Fund Program in 1995, Vermont's legislature recognized that the state's economic prosperity depends on maintaining "Vermont's unique environmental image" and that "the function of state policy and of the policies of our existing educational institutions provides an opportunity to position the state as a primary sustainable economic development educational center." Maine's legislature established a small enterprise growth fund to offer "patient" sources of venture capital to green or socially motivated businesses in 1997. The state has also made public investments for clean technology research since 2007. The state of New Hampshire's interest also dates to 2007, with its climate change action plan.

31 Anair and Mahmassani, *Clean Energy Momentum.*
32 Blittersdorf, "VT Could Lead in Wind Power."
33 Coley and Hess, "Green Energy Laws and Republican Legislators in the United States."
34 Coley and Hess, "Green Energy Laws and Republican Legislators in the United States."
35 Geels, "Regime Resistance against Low-Carbon Transitions," 27.
36 Geels, "Regime Resistance against Low-Carbon Transitions," 26.
37 Reicher, "Discuss Wind Power Based on Facts."
38 Ballek, "Commentary."
39 Totten, "Tilting at Turbines."
40 "PUC Chairman Took Equity Stake in Wind Company."
41 Schalit, "LD 1750."
42 Maynard, "Apparently, Greed Is OK When It Is 'Green.'"
43 "Opinion: Green Love Is Blind."
44 Aklin and Urpelainen, *Renewables.*
45 Polhamus, "Program to Promote Renewable Energy Gets New Scrutiny."
46 Gribkoff, "Regulators Approve GMP Rate Request after Public Criticism."
47 Dobbs, "Report Confirms GMP Customers Not Well-Served by Alternative Regulation."
48 Angwin, *Shorting the Grid.*
49 Pazniokas, "Governors Want Sunlight on Secretive ISO-New England."
50 Hughes, *Networks of Power.*
51 Beder, *Power Play.*
52 Beder, *Power Play.*
53 Grimley, "Just How Democratic Are Rural Electric Cooperatives?"
54 Zehner, *Green Illusions.*
55 Angwin, *Shorting the Grid,* 147.
56 Morehouse, "NERA Counters Broad Opposition to FERC Net Metering Petition."

Chapter 6   Scripted in Chaos

1 Lynch, *Switch.*
2 Podobnik, *Global Energy Shifts.*
3 Hess, "Sustainability Transitions," 279.
4 Harjanne and Korhonen, "Abandoning the Concept of Renewable Energy."
5 Worland, "Why Fossil Fuel Companies Are Reckoning with Climate Change."

6 Thanks to the work of nineteenth-century British economist William Stanley Jevons, we have long acknowledged that fossil fuels' profitability can decline even as scarcity worsens. Based on the depletion of British coal reserves, the Jevons paradox holds that scarce resources will sell for prices that are counterintuitively low as extraction technology improves and competition compels firms to lower their prices, accelerating consumer demand even more.

7 Rohracher, "Managing the Technological Transition to Sustainable Construction of Buildings."

8 Rohracher, "Energy Systems in Transition."

9 Rohracher, "Energy Systems in Transition."

10 Gold, *The Boom.*

11 Harjanne and Korhonen, "Abandoning the Concept of Renewable Energy."

12 Mingle, "The Methane Detectives."

13 Hmiel et al., "Preindustrial 14CH4 Indicates Greater Anthropogenic Fossil CH4 Emissions."

14 Pandey et al., "Satellite Observations Reveal Extreme Methane Leakage."

15 Galbraith and Price, *The Great Texas Wind Rush.*

16 Weis, "As Climate Change Threatens Earth, US to Open Nearly 200 Power Plants."

17 Krauss, "Exxon Mobil Tripling Its Bet on the Hottest U.S. Shale Field"; Olson, "Big Oil Companies Reap Windfall from Ethanol Rules."

18 Unruh, "Understanding Carbon Lock-In."

19 Orsted, "Wind Power."

20 Ismay, "Commentary."

21 Marks et al., "Vertical Market Power in Interconnected Natural Gas and Electricity Markets."

22 Angwin, *Shorting the Grid.*

23 Crowley, "Subsidies a Bone of Contention as Renewable Energy Producers Seek Federal Ruling."

24 Angwin, *Shorting the Grid,* 184.

25 Geels, "Regime Resistance against Low-Carbon Transitions," 27.

26 Geels, "Regime Resistance against Low-Carbon Transitions," 34.

27 Stirling, "Multicriteria Diversity Analysis."

28 Geels, "Regime Resistance against Low-Carbon Transitions," 35.

29 Jenkins et al., "Energy Justice."

30 Newell and Mulvaney, "The Political Economy of the 'Just Transition.'"

31 Jenkins et al., "Energy Justice."

32 Burke and Stephens, "Political Power and Renewable Energy Futures."

33 Malin, *The Price of Nuclear Power.*

34 Manning, "The Oil We Eat."

35 Olson, "Big Oil Companies Reap Windfall from Ethanol Rules."

36 Fletcher, *Bottled Lightning.*

37 Wright and Nyberg, *Climate Change, Capitalism, and Corporations.*

38 Goldstein, *Planetary Improvement,* 157.

39 Goldstein, *Planetary Improvement,* 157.

40 Goldstein, *Planetary Improvement,* 158.

Chapter 7   Why We Follow the Slow Transition Road Map

1 Geels, "Regime Resistance against Low-Carbon Transitions."
2 Rooney, "Knowledge, Economy, Technology and Society."
3 Franta and Supran, "The Fossil Fuel Industry's Invisible Colonization of Academia."
4 Wertz, "Reading, Writing and Fracking?"
5 Beder, *Power Play.*
6 Harris, "Shell Has a Plan to Profit from Climate Change."
7 Zehner, "Conjuring Clean Energy."
8 Zehner, "Conjuring Clean Energy."
9 Gibbs, *Planet of the Humans.*
10 Gramling, "What Michael Moore's New Film Gets Wrong about Renewable Energy."
11 "Fact Check Bible."
12 Whitlock, "Just Plain Wrong."
13 Harjanne and Korhonen, "Abandoning the Concept of Renewable Energy"; York and Bell, "Energy Transitions or Additions?"
14 Westervelt, "Drilled."
15 Castoulis, "'Switch' Explores a World of Fuel Options."
16 Sharkey, "Review."
17 Wright and Nyberg, *Climate Change, Capitalism, and Corporations.*
18 Szasz, *Shopping Our Way to Safety.*
19 Nicholson-Cole, "'Fear Won't Do It'"; Slocum, "Polar Bears and Energy-Efficient Lightbulbs."
20 Zehner, *Green Illusions,* 149.
21 Wilson and Stephens, "Wind Deployment in the United States."
22 Jasanoff, *States of Knowledge*; Latour, *Reassembling the Social.*
23 Haraway, "Situated Knowledges"; Knorr-Cetina, *Epistemic Cultures.*
24 Ohanian, "Carbon Myopia in Montpelier."
25 Intermittent, volatile wind is inferior to strong and steady wind in terms of capacity to offset fossil fuel electricity sources. This is the reason that turbines in the central plains and along windy coastlines work well.
26 Sovacool et al., "Valuing the Manufacturing Externalities of Wind Energy."
27 Mills, "If You Want 'Renewable Energy,' Get Ready to Dig."
28 Goldstein, *Planetary Improvement.*
29 Goldstein, *Planetary Improvement,* 91.
30 Olson-Hazboun, "'Why Are We Being Punished and They Are Being Rewarded?'"
31 Macdonald, "Robotic Inspectors Developed to Fix Wind Farms."
32 Bakke, *The Grid.*

Chapter 8   Ecological Modernizations or Capitalist Treadmills?

1 Mol, "The Refinement of Production."
2 Mol and Janicke, "The Origins and Theoretical Foundations of Ecological Modernisation Theory."
3 Mol and Spaargaren, "Ecological Modernization and the Environmental State."

4   Buttel, "Ecological Modernization as Social Theory."
5   Christoff, "Ecological Modernisation, Ecological Modernities."
6   Initially, funding for the CEDF derived from annual payments negotiated from Vermont's nuclear power plant in exchange for allowing the plant's spent fuel rods to be stored on site. Secondary federal funding has also come from the State Energy Program (SEP) and the Energy Efficiency and Conservation Block Grant (EECBG) program. Permanent closure of the nuclear facility in 2014 has limited the fund's resources. However, state appropriations and now provisions for noncompliance under the state's 2016 energy plan are projected to divert funds from the private sector into the CEDF.
7   Cardwell, "Utility Helps Wean Vermonters from the Electric Grid."
8   Hess, "A Political Economy of Sustainability," 251.
9   Toke, *Ecological Modernisation and Renewable Energy*.
10 Hess, "A Political Economy of Sustainability," 244.
11 Schnaiberg, *The Environment*.
12 Bell, *An Invitation to Environmental Sociology*.
13 Prasanna et al., "Recent Experiences with Tariffs for Saving Electricity in Households."
14 Shove, "Efficiency and Consumption."
15 Mol, Spaargaren, and Sonnenfeld, "Ecological Modernisation Theory."
16 Christoff, "Ecological Modernisation, Ecological Modernities."
17 Buttel, "Ecological Modernization as Social Theory."
18 Therrien, "Teflon Town."
19 Geels, "Regime Resistance against Low-Carbon Transitions."
20 Trombly, "Canadians Ask for Longer Ban on Leachate Treatment in Memphremagog"; Blazer, "Prevalence of Malignant Melanoma in Brown Bullhead from Lake Memphremagog Greater than Expected."
21 Gokee, "Chittenden Solid Waste District to Pay $400,000 to Settle Glass Dumping Case."
22 MacNeil and Paterson, "Neoliberal Climate Policy."

Chapter 9   Energy and "Justice" in the Mountains

1   Klein, *This Changes Everything*.
2   Harrison, "Neoliberal Environmental Justice."
3   Harrison, "Neoliberal Environmental Justice."
4   Sovacool and Dworkin, "Energy Justice."
5   Krogh, "Industrial Wind Turbine Development and Loss of Social Justice?"
6   Genter, "Falmouth Ordered to Shut Down Turbines."
7   The one farmer in the region hoping to host a single grid-scale wind turbine has met with fierce opposition from neighbors and antiwind groups.
8   Anhoury, "N.Y.C. Consulting with Quebec's Indigenous Peoples before Inking Hydro Deal."
9   Evans-Brown and McCarthy, "Powerline—Outside/In."
10 Hongoltz-Hetling, "New England's Drive for Canadian Hydropower Threatens Native Population's Way of Life."
11 Mills, "If You Want 'Renewable Energy,' Get Ready to Dig."
12 USGS National Minerals Information Center, "Rare Earths."

13 Lehr, "Addressing Forced Labor in the Xinjiang Uyghur Autonomous Region."
14 Bell, "Environmental Injustice and the Pursuit of a Post-Carbon World."
15 Mulvaney, *Solar Power: Innovation, Sustainability, and Environmental Justice.*
16 Hornborg, "Zero-Sum World."
17 Foster, *The Vulnerable Planet.*
18 Park and Pellow, *The Slums of Aspen.*
19 McMichael, *Development and Social Change.*
20 Gilio-Whitaker, *As Long as the Grass Grows*, 92.
21 Grossman, "Unlikely Alliances."
22 Nadasdy, *Hunters and Bureaucrats.*
23 Krech, *The Ecological Indian.*
24 Dunlap, "The 'Solution' Is Now the 'Problem.'"
25 Smith, *Decolonizing Methodologies.*
26 Vanderbeck, "Vermont and the Imaginative Geographies of American Whiteness."
27 Bonilla-Silva, *Racism without Racists.*
28 Evancie, "What's the History of the Underground Railroad in Vermont?"
29 Whitfield, *The Problem of Slavery in Early Vermont, 1777–1810.*
30 Anderson, *The American Census.*
31 "Vt. Officials Push Diversity after SNL Skit on Whiteness."
32 Corwin, "Remembering Vermont's 19th Century Black Communities."
33 Corwin, "'Laugh because It's Funny . . . Cry because It's True.'"

Chapter 10 Reimagining Energy

1 Podobnik, *Global Energy Shifts.*
2 Devine-Wright, "Energy Citizenship."
3 Devine-Wright, "Energy Citizenship."
4 Burke and Stephens, "Political Power and Renewable Energy Futures," 79.
5 Jasanoff, "Just Transitions."
6 Harrabin, "Pollution Warning over Car Tyre and Brake Dust"; Harrabin, "Electric Cars 'Will Not Solve Transport Problem,' Report Warns."
7 Harjanne and Korhonen, "Abandoning the Concept of Renewable Energy," 337.
8 Zehner, *Green Illusions.*
9 Bowden, "'Life. Brought to You by' . . . Coal?"
10 Shove, "Efficiency and Consumption."
11 Hess, "A Political Economy of Sustainability," 20.
12 McGrath, "Climate Change."
13 Geels, "Regime Resistance against Low-Carbon Transitions," 37.
14 Hess, "Electricity Transformed."
15 Cameron and Hicks, "Performative Research for a Climate Politics of Hope."
16 Geels, "Regime Resistance against Low-Carbon Transitions," 17.
17 Brooks, "Microgrids Seek Path out of Regulatory Limbo."
18 Mims, "Why Your Next Home Might Not Need Any Energy at All."
19 Gillis, "Opinion."

Epilogue

1 Hirschkorn, "Big Money from Away Flows in Rival Maine Power Line Campaigns."

Bibliography

Aklin, Michaël, and Johannes Urpelainen. *Renewables: The Politics of a Global Energy Transition*. Cambridge, MA: MIT Press, 2018.

Albers, Jan. *Hands on the Land: A History of the Vermont Landscape*. Cambridge, MA: MIT Press, 2000.

Anair, Don, and Amine Mahmassani. *Clean Energy Momentum: Ranking State Progress*. Union of Concerned Scientists, 2017.

Angwin, Meredith. *Shorting the Grid: The Hidden Fragility of Our Electric Grid*. Wilder, VT: Carnot Communications, 2000.

Anderson, Dave. "Utilities, Koch Groups Back Sununu's Veto of Net Metering Bill in NH." Energy and Policy Institute, August 11, 2018.

Anderson, Margo J. *The American Census: A Social History*. New Haven, CT: Yale University Press, 2015.

Anhoury, Mia. "N.Y.C. Consulting with Quebec's Indigenous Peoples before Inking Hydro Deal." *Montreal Gazette*, July 24, 2019.

Associated Press. "Vt. Officials Push Diversity after SNL Skit on Whiteness." October 2, 2018.

Bakke, Gretchen. *The Grid: The Fraying Wires between Americans and Our Energy Future*. New York: Bloomsbury Publishing, 2016.

Ballek, Keith. "Commentary: A Question of VPIRG's Conflicts." *VT Digger*, October 19, 2018.

Batel, Susana, and Patrick Devine-Wright. "A Critical and Empirical Analysis of the National-Local 'Gap' in Public Responses to Large-Scale Energy Infrastructures." *Journal of Environmental Planning and Management* 58, no. 6 (2014): 1076–1095.

Beale, Calvin L. "A Further Look at Nonmetropolitan Population Growth since 1970." *American Journal of Agricultural Economics* 58, no. 5 (December 1976): 953–958.

Beck, Ulrich. "Climate for Change, or How to Create a Green Modernity?" *Theory, Culture & Society* 27, no. 2–3 (2010): 254–266.

———. "On the Way to the Industrial Risk-Society? Outline of an Argument." *Thesis Eleven* 23, no. 1 (1989): 86–103.

————. "Risk Society and the Provident State." In *Risk, Environment and Modernity: Towards a New Ecology*, edited by Scott Lash, Bronislaw Szerszynski, and Brian Wynne, 27–43. London: Sage Publications, 1996.

Beckley, Thomas M. "Energy and the Rural Sociological Imagination." *Journal of Rural Social Sciences* 32, no. 2 (2017): 69–97.

Beder, Sharon. *Power Play: The Fight to Control the World's Electricity*. New York: The New Press, 2003.

Bell, Michael Mayerfeld. *An Invitation to Environmental Sociology*, 4th ed. Los Angeles: Sage, 2012.

Bell, Shannon Elizabeth. "Environmental Injustice and the Pursuit of a Post-Carbon World: The Unintended Consequences of the Clean Air Act as a Cautionary Tale for Solar Energy Development." *Brooklyn Law Review* 82, no. 2 (2017): 529–557.

Benford, Robert D., and David A. Snow. "Framing Processes and Social Movements: An Overview and Assessment." *Annual Review of Sociology* 26 (2000): 611–639.

Blittersdorf, David. "VT Could Lead in Wind Power." *Burlington Free Press*, October 19, 2002.

Bonilla-Silva, Eduardo. *Racism without Racists: Color-Blind Racism and the Persistence of Racial Inequality in the United States*. Lanham, MD: Rowman & Littlefield, 2006.

Bowden, Vanessa. "'Life. Brought to You by' . . . Coal? Business Responses to Climate Change in the Hunter Valley, NSW, Australia." *Environmental Sociology* 4, no. 2 (April 3, 2018): 275–285.

Brooks, Michael. "Microgrids Seek Path out of Regulatory Limbo." *RTO Insider*, November 2018.

Brown, Dona. *Back to the Land: The Enduring Dream of Self-Sufficiency in Modern America*. Madison: University of Wisconsin Press, 2011.

————. *Inventing New England: Regional Tourism in the 19th Century*. Washington, DC: Smithsonian Books, 1997.

Burke, Matthew J., and Jennie C. Stephens. "Political Power and Renewable Energy Futures: A Critical Review." *Energy Research & Social Science* 35 (January 1, 2018): 78–93.

Burlington Free Press. "Big Wind, Small State." Editorial. March 13, 2005.

Buttel, Frederick H. "Ecological Modernization as Social Theory." *Geoforum* 31, no. 1 (2000): 57–65.

Cameron, Jenny, and Jarra Hicks. "Performative Research for a Climate Politics of Hope: Rethinking Geographic Scale, 'Impact' Scale, and Markets." *Antipode* 46, no. 1 (2014): 53–71.

Cardwell, Diane. "Utility Helps Wean Vermonters from the Electric Grid." *New York Times*, July 29, 2017.

Carlsen, Audrey. "Could Wind Turbines Be Toxic to the Ear?" National Public Radio, April 2, 2013.

Castleton University Polling Institute. "Castleton Poll Examines Vermonters' Views on Wind Power: Complete Poll Results." February, 26 2013.

Castoulis, Jeanette. "'Switch' Explores a World of Fuel Options." *New York Times*, September 20, 2012.

Chapman, Simon, Alexis St. George, Karen Waller, and Vince Cakic. "Spatio-Temporal Differences in the History of Health and Noise Complaints about Australian Wind Farms: Evidence for the Psychogenic, and 'Communicated

Disease' Hypothesis," Unpublished Manuscript. 2013. Accessed January 15, 2021. https://ses.library.usyd.edu.au/bitstream/handle/2123/8977/Complaints%20 FINAL.pdf?sequence=4&isAllowed=y

Chen, Pauline. "Teaching Doctors about Food and Diet." *New York Times*, September 16, 2010.

Christoff, Peter. "Ecological Modernisation, Ecological Modernities." *Environmental Politics* 5, no. 3 (1996): 476–500.

Coley, Jonathan S., and David J. Hess. "Green Energy Laws and Republican Legislators in the United States." *Energy Policy* 48, no. 1 (2012): 576–583.

Corwin, Emily. "'Laugh because It's Funny . . . Cry because It's True': SNL Sketch Skewers Lack of Diversity in Vt." Vermont Public Radio, October 2, 2018.

———. "Remembering Vermont's 19th Century Black Communities." *Brave Little State*. Vermont Public Radio, June 2020.

Crichton, Fiona, and Keith J. Petrie. "Health Complaints and Wind Turbines: The Efficacy of Explaining the Nocebo Response to Reduce Symptom Reporting." *Environmental Research* 140 (2015): 449–455.

Cronon, William. *Changes in the Land: Indians, Colonists, and the Ecology of New England*. New York: Hill and Wang, 2003.

Crowley, Brendan. "Subsidies a Bone of Contention as Renewable Energy Producers Seek Federal Ruling." *CT Examiner*, January 7, 2021.

Devine-Wright, Patrick. "Beyond NIMBYism: Towards an Integrated Framework for Understanding Public Perceptions of Wind Energy." *Wind Energy* 8, no. 2 (2005): 125–139.

———. "Energy Citizenship: Psychological Aspects of Evolution in Sustainable Energy Technologies." In *Governing Technology for Sustainability*, edited by Joseph Murphy, 63–86. London: Earthscan, 2007.

———. "From Backyards to Places: Public Engagement and the Emplacement of Renewable Energy Technologies." In *Renewable Energy and the Public: From NIMBY to Participation*, edited by Patrick Devine-Wright, 57–70. London: Earthscan, 2011.

———. "Public Engagement with Large-Scale Renewable Energy Technologies: Breaking the Cycle of NIMBYism." *Wiley Interdisciplinary Reviews: Climate Change* 2, no. 1 (2011): 19–26.

———. "Think Global, Act Local? The Relevance of Place Attachments and Place Identities in a Climate Changed World." *Global Environmental Change* 23, no. 1 (2013): 61–69.

De Wit, Wijnand, and Nathan Bigaud. *No Plastic in Nature: Assessing Plastic Ingestion from Nature to People*. Gland, Switzerland: World Wide Fund for Nature, 2019.

Dickinson, Tim. "How Big Oil and Big Soda Kept the Plastics Crisis a Secret for Decades." *Rolling Stone*, March 3, 2020.

Dillon, John. "Activists Press for Close Review of Vermont Utilities' Canadian Ownership." Vermont Public Radio, 2019.

———. "Transmission Grid Bottlenecks in Northeast Kingdom Stall Solar Development." Vermont Public Radio, 2020.

———. "Vermont's Energy Credits Questioned, Renewable Program Likely to Change." Vermont Public Radio, 2015.

Dobbs, Taylor. "Report Confirms GMP Customers Not Well-Served by Alternative Regulation." Vermont Public Radio, 2017.

Dunlap, Alexander. "The 'Solution' Is Now the 'Problem': Wind Energy, Colonisation and the 'Genocide-Ecocide Nexus' in the Isthmus of Tehuantepec, Oaxaca." *International Journal of Human Rights*, November 17, 2017, 1–24.

Evancie, Angela. "What's the History of the Underground Railroad in Vermont?" *Brave Little State*. Vermont Public Radio, October 2017.

Evans-Brown, Sam, and Hannah McCarthy. "Powerline—Outside/In." New Hampshire Public Radio, 2017.

"Fact Check Bible." *Planet of the Humans*. Accessed December 14, 2020. https://planetofthehumans.com/fact-check-bible/.

Farrell, John. *Democratizing the Electricity System: A Vision for the 21st Century Grid*. Minneapolis, MN: The New Rules Project, 2011.

Firetrace International. *In the Line of Fire*. 2020. Accessed January 14, 2021. https://www.firetrace.com/hubfs/_img/reports/Firetrace-Report-In-The-Line -Of-Fire.pdf.

Fletcher, Seth. *Bottled Lightning: Superbatteries, Electric Cars, and the New Lithium Economy*. New York: Hill and Wang, 2011.

Foster, John Bellamy. *The Vulnerable Planet: A Short Economic History of the Environment*. New York: New York University Press, 1999.

Foucault, Michel. *Archaeology of Knowledge*. Abingdon, UK: Routledge, 2013.

Franta, Benjamin, and Geoffrey Supran. "The Fossil Fuel Industry's Invisible Colonization of Academia." *Guardian*, March 17, 2017.

Freudenburg, William R., and Susan K. Pastor. "NIMBYs and LULUs: Stalking the Syndromes." *Journal of Social Issues* 48, no. 4 (January 1, 1992): 39–61.

———. "Public Responses to Technological Risks: Toward a Sociological Perspective." *Sociological Quarterly* 33, no. 3 (1992): 389–412.

Galbraith, Kate (Catherine), and Asher Price. *The Great Texas Wind Rush: How George Bush, Ann Richards, and a Bunch of Tinkerers Helped the Oil and Gas State Win the Race to Wind Power*. Austin: University of Texas Press, 2013.

Gaventa, John. *Power and Powerlessness: Quiescence and Rebellion in an Appalachian Valley*. Champaign: University of Illinois Press, 1982.

Geels, F. W. "Regime Resistance against Low-Carbon Transitions: Introducing Politics and Power into the Multi-Level Perspective." *Theory, Culture & Society*, 31, no. 5 (2014): 21–40.

Genter, Ethan. "Falmouth Ordered to Shut Down Turbines." *Cape Cod Times*, June 20, 2017.

Gibbs, Jeff, dir. *Planet of the Humans*. Rumble Media, 2020.

Gilio-Whitaker, Dina. *As Long as Grass Grows: The Indigenous Fight for Environmental Justice, from Colonization to Standing Rock*. Boston: Beacon Press, 2019.

Gillis, Justin. "Opinion: An Important Vote for the Climate." *New York Times*, March 20, 2019.

Gokee, Amadnda. "Vermont Regulators Cut Incentives to Switch to Solar Energy." *VTDigger*, December 20, 2020.

Gold, Russell. *The Boom: How Fracking Ignited Mills the American Energy Revolution and Changed the World*. New York: Simon & Schuster, 2014.

Golding, Shaun A. "Rural Gentrification in the United States 1970–2000: A Demographic Analysis of Its Footprint, Impacts, and Implications." PhD diss., University of Wisconsin–Madison, 2012.

Goldman, Hal. "James Taylor's Progressive Vision: The Green Mountain Parkway." *Vermont History* 63, no. 3 (1995): 158–179.

Goldstein, Jesse. *Planetary Improvement: Cleantech Entrepreneurship and the Contradictions of Green Capitalism.* Cambridge, MA: MIT Press, 2018.

Grady, Denise. "Physician Revives a Dying Art: The Physical." *New York Times,* October 11, 2010.

Gramling, Carolyn. "What Michael Moore's New Film Gets Wrong about Renewable Energy." *Science News,* May 11, 2020. Accessed June 29, 2020. https://www.science news.org/article/what-michael-moore-new-film-gets-wrong-about-renewable -energy.

Gramsci, Antonio, and Quintin Hoare. *Selections from the Prison Notebooks.* London: Lawrence & Wishart, 1971.

Gribkoff, Elizabeth. "Regulators Approve GMP Rate Request after Public Criticism." *VT Digger,* December 28, 2018.

Grimley, Matt. "Just How Democratic Are Rural Electric Cooperatives?" Institute for Local Self-Reliance, January 13, 2016.

Grossman, Zoltan. "Unlikely Alliances: Treaty Conflicts and Environmental Cooperation between Native American and Rural White Communities." *American Indian Culture and Research Journal* 29, no. 4 (2005): 21–43.

Guy, Simon, and Elizabeth Shove. *The Sociology of Energy, Buildings, and the Environment: Constructing Knowledge, Designing Practice.* Abingdon, UK: Routledge, 2014.

Hager, Bradford H. "Commentary: Hydro-Québec Offers Misleading Claims about Power's Climate Impact." *Portland Press Herald,* December 5, 2019.

Hager, Carol, and Mary Alice Haddad. "A New Look at NIMBY." In *NIMBY Is Beautiful: Cases of Local Activism and Environmental Innovation around the World,* 1st ed., edited by Carol Hager and Mary Alice Haddad, 1–14. New York: Berghahn Books, 2015.

Hajer, Maarten A. *The Politics of Environmental Discourse: Ecological Modernization and the Policy Process.* Oxford: Clarendon Press, 1995.

Halfacree, K. "Contrasting Roles for Post Productivist Countryside." In *Contested Countryside Cultures,* edited by Paul Cloke and Jo Little, 70–93. London: Routledge, 1997.

Haraway, Donna. "Situated Knowledges: The Science Question in Feminism and the Privilege of Partial Perspective." *Feminist Studies* 14, no. 3 (1988): 575–599.

Harjanne, A., and J. M. Korhonen. "Abandoning the Concept of Renewable Energy." *Energy Policy* 127 (2019): 330–340.

Harrabin, Roger. "Electric Cars 'Will Not Solve Transport Problem,' Report Warns." *BBC News,* July 5, 2019. Accessed February 17, 2020. https://www.bbc.com/news /uk-48875361.

———. "Pollution Warning over Car Tyre and Brake Dust." *BBC News,* July 11, 2019. Accessed February 17, 2020. https://www.bbc.com/news/business-48944561.

Harris, Malcolm. "Shell Has a Plan to Profit from Climate Change." *New York Magazine,* March 3, 2020.

Harrison, Blake A. *The View from Vermont: Tourism and the Making of an American Rural Landscape.* Lebanon, NH: UPNE, 2006.

Harrison, Jill Lindsey. "Neoliberal Environmental Justice: Mainstream Ideas of Justice in Political Conflict over Agricultural Pesticides in the United States." *Environmental Politics* 23, no. 4 (2014): 650–669.

Harvey, David. *A Brief History of Neoliberalism*. New York: Oxford University Press, 2007.

Herrick, John. "Green Businesses Press for Carbon Tax." *VT Digger*, April 8, 2015.

Hess, David J. "A Political Economy of Sustainability: Alternative Pathways and Industrial Innovation." In *Pragmatic Sustainability: Theoretical and Practical Tools*, edited by Steven Moore, 236–258. Cambridge, MA: MIT Press, 2010.

———. "Electricity Transformed: Neoliberalism and Local Energy in the United States." *Antipode* 43, no. 3 (2011): 1056–1057.

———. "Sustainability Transitions: A Political Coalition Perspective." *Research Policy* 43, no. 2 (2014): 278–283.

———. "The Green Transition, Neoliberalism, and the Technosciences." In *Neoliberalism and Technoscience: Critical Assessments*, edited by Marja Ylönen, 209. Abingdon, UK: Routledge, 2016.

Hirschkorn, Phil. "Big Money from Away Flows in Rival Maine Power Line Campaigns." WMTW. December 18, 2020. Accessed January 14, 2021. https://www.wmtw.com/article/big-money-from-away-flows-in-rival-maine-power-line-campaigns/35017979.

Hmiel, Benjamin, V. V. Petrenko, M. N. Dyonisius, C. Buizert, A. M. Smith, P. F. Place, C. Harth, R. Beaudette, Quan Hua, B. Yang, I. Vimont, S. E. Michel, Jeff Severinghaus, David Etheridge, Tony Bromley, Jochen Schmitt, Xavier Fain, Ray F. Weiss, and E. Dlugokencky. "Preindustrial 14CH4 Indicates Greater Anthropogenic Fossil CH_4 Emissions." *Nature* 578, no. 7795 (February 19, 2020): 409–412.

Holland, Robert R., and Benjamin Luce. "The Cost-Effectiveness of 21 Vestas V112 Wind Turbines on the Lowell Mountains Ridgeline." 2011.

Hongoltz-Hetling, Matt. "New England's Drive for Canadian Hydropower Threatens Native Population's Way of Life." *Valley News*, November 30, 2019.

Hornborg, Alf. "Zero-Sum World: Challenges in Conceptualizing Environmental Load Displacement and Ecologically Unequal Exchange in the World-System." *International Journal of Comparative Sociology* 50, no. 3–4 (June 20, 2009): 237–262.

Hughes, Thomas Parke. *Networks of Power: Electrification in Western Society, 1880–1930*. Baltimore: Johns Hopkins University Press, 1983.

Ismay, David. "Commentary: ISO's 'Sky Is Falling' Message." *VTDigger*, June 18, 2018.

Jacob, Jeffrey. *New Pioneers: The Back-to-the-Land Movement and the Search for a Sustainable Future*. University Park: Pennsylvania State University Press, 2010.

Jacquet, Jeffrey B. "Landowner Attitudes toward Natural Gas and Wind Farm Development in Northern Pennsylvania." *Energy Policy* 50 (2012): 677–688.

Jasanoff, Sheila. "Just Transitions: A Humble Approach to Global Energy Futures." *Energy Research & Social Science* 35 (January 1, 2018): 11–14.

———. *States of Knowledge: The Co-production of Science and the Social Order*. Abingdon, UK: Routledge, 2004.

Jenkins, Kirsten, Darren McCauley, Raphael Heffron, Hannes Stephan, and Robert Rehner. "Energy Justice: A Conceptual Review." *Energy Research & Social Science* 11 (January 1, 2016): 174–182.

Jones, Kevin B., Mark James, and Heather Huebner. "Do You Know Who Owns Your Solar Energy? The Growing Practice of Separating Renewable Attributes from

Renewable Energy Development and Its Impact on Meeting Our Climate Goals." *Fordham Environmental Law Review* 28, no. 2 (2017): 197–241.

Judd, Richard W. *Common Lands, Common People: The Origins of Conservation in Northern New England.* Cambridge, MA: Harvard University Press, 1997.

Kacprzynski, F. T., and Alan R. Graefe. *Proceedings of the 1990 Northeastern Recreation Research Symposium.* US Department of Agriculture, Forest Service, Northeastern Forest Experiment Station, General Technical Report NE-145, 1990.

Kallman, Meghan Elizabeth, and Scott Frickel. "Power to the People: Industrial Transition Movements and Energy Populism." *Environmental Sociology* 5, no. 3 (July 3, 2019): 255–268.

Kantamneni, Abhilash, Richelle Winkler, Lucia Gauchia, and Joshua M. Pearce. "Emerging Economic Viability of Grid Defection in a Northern Climate Using Solar Hybrid Systems." *Energy Policy* 95 (August 2016): 378–389.

Kelly, David F. "Opinion: An Alternative Form of Renewable Energy." *VTDigger*, October 25, 2020.

Klein, Naomi. *This Changes Everything: Capitalism vs. the Climate.* New York: Simon & Schuster, 2015.

Knight, Pat, and Jamie Hall. *Wholesale Cost Savings of Distributed Solar in New England.* Prepared for SunCommon. Synapse Energy Economics, August 28, 2018.

Knorr-Cetina, Karin. *Epistemic Cultures: How the Sciences Make Knowledge.* Cambridge, MA: Harvard University Press, 2009.

Krauss, Clifford. "Exxon Mobil Tripling Its Bet on the Hottest U.S. Shale Field." *New York Times*, January 30, 2018.

Krech, Shepard. *The Ecological Indian: Myth and History.* New York: W.W. Norton & Company, 2000.

Krogh, C. M. E. "Industrial Wind Turbine Development and Loss of Social Justice?" *Bulletin of Science, Technology & Society* 31, no. 4 (July 19, 2011): 321–333.

Kuentzel, W. F, and V. M Ramaswamy. "Tourism and Amenity Migration: A Longitudinal Analysis." *Annals of Tourism Research* 32, no. 2 (2005): 419–438.

Lachapelle, Paul, and Stephen McCool. "Exploring the Concept of 'Ownership' in Natural Resource Planning." *Society & Natural Resources* 18, no. 3 (February 2005): 279–285.

Latour, Bruno. *Reassembling the Social: An Introduction to Actor-Network-Theory.* New York: Oxford University Press, 2005.

———. *We Have Never Been Modern.* Cambridge, MA: Harvard University Press, 2012.

Lehr, Amy K. *Addressing Forced Labor in the Xinjiang Uyghur Autonomous Region: Toward a Shared Agenda.* Center for Strategic and International Studies. 2020.

Levy, David L., and Peter J. Newell. "Business Strategy and International Environmental Governance: Toward a Neo-Gramscian Synthesis." *Global Environmental Politics* 2, no. 4 (November 2002): 84–101.

Li, Yan, Eugenia Kalnay, Safa Motesharrei, Jorge Rivas, Fred Kucharski, Daniel Kirk-Davidoff, Eviatar Bach, and Ning Zeng. "Climate Model Shows Large-Scale Wind and Solar Farms in the Sahara Increase Rain and Vegetation." *Science* 361, no. 6406 (September 7, 2018): 1019–1022.

Lloyd, Graham. "Wind Farm Advocate Simon Chapman Sorry for False Allegations." *Australian*, August 19, 2015.

Luce, Ben. "Issues with Wind in Our Region?" Accessed October 1, 2018. http://www
.nhwindwatch.org/Issues_WindPowerOnNHRidgelines.pdf.
Lukes, Steven. *Power.* New York: New York University Press, 1986.
Lynch, Harry, dir. *Switch.* Arcos Films, 2012.
Macdonald, Ken. "Robotic Inspectors Developed to Fix Wind Farms." *BBC News,*
October 14, 2019.
MacNeil, Robert, and Matthew Paterson. "Neoliberal Climate Policy: From Market
Fetishism to the Developmental State." *Environmental Politics* 21, no. 2 (2012):
230–247.
Malin, Stephanie A. *The Price of Nuclear Power: Uranium Communities and Environ-
mental Justice.* New Brunswick, NJ: Rutgers University Press, 2015.
Manning, Richard. "The Oil We Eat: Following the Food Chain Back to Iraq."
Harper's Magazine, May 2004.
Marcouiller, D. W., K. K. Kim, and S. C. Deller. "Natural Amenities, Tourism and
Income Distribution." *Annals of Tourism Research* 31, no. 4 (2004): 1031–1050.
Marks, Levi, Charles F. Mason, Kristina Mohlin, and Matthew Zaragoza-Watkins.
"Vertical Market Power in Interconnected Natural Gas and Electricity Markets."
CESifo Working Paper series no. 6687, November 17, 2017.
Maynard, Robert. "Apparently, Greed Is OK When It Is 'Green.'" *True North Reports,*
2015.
McClymont, K., and P. O'Hare. "'We're Not NIMBYs!' Contrasting Local Protest
Groups with Idealised Conceptions of Sustainable Communities." *Local Environ-
ment: The International Journal of Justice and Sustainability* 13, no. 4 (June 2008):
321–335.
McGrath, Matt. "Climate Change: Electrical Industry's 'Dirty Secret' Boosts
Warming." *BBC News,* September 13, 2019. Accessed February 4, 2020. https://
www.bbc.com/news/science-environment-49567197.
McGregor, Natalie. "Australian Court Links Wind Turbine Noise with Possible
Diseases." *Hamilton Spectator,* February 15, 2018.
McMichael, Philip. *Development and Social Change: A Global Perspective.* Thousand
Oaks, CA: Pine Forge Press, 2007.
Miller, Lee M., and David W. Keith. "Climatic Impacts of Wind Power." *Joule* 2,
no. 12 (December 19, 2018): 2618–2632.
———. "Observation-Based Solar and Wind Power Capacity Factors and Power
Densities." *Environmental Research Letters* 13, no. 10 (October 4, 2018): 104008.
Mills, Mark P. "If You Want 'Renewable Energy,' Get Ready to Dig." *Wall Street
Journal,* August 5, 2019.
Mills, Paul. "The Unheralded Champion of Maine's 'Vacationland' License Plate
Motto." *Daily Bulldog,* July 9, 2014.
Mims, Christopher. "Why Your Next Home Might Not Need Any Energy at All."
Wall Street Journal, December 22, 2018.
Mingle, Johnathan. "The Methane Detectives: On the Trail of a Global Warming
Mystery." *Undark.* Accessed January 15, 2021. https://undark.org/2019/05/13
/methane-global-warming-climate-change-mystery/.
Mirowski, Philip. "Neoliberalism: The Movement That Dare Not Speak Its Name."
American Affairs 2, no. 1 (Spring 2018): 118–141.
Mol, Arthur P. J. *The Refinement of Production: Ecological Modernization Theory and
the Chemical Industry.* Utrecht, Netherlands: Jan van Arkel Books, 1995.

Mol, Arthur P. J., and Gert Spaargaren. "Ecological Modernization and the Environmental State." In *The Environmental State under Pressure*, edited by Arthur P. J. Mol and Frederick H. Buttel, 33–52. Bingley, UK: Emerald Group Publishing Limited, 2002.

Mol, Arthur P. J., and Martin Janicke. "The Origins and Theoretical Foundations of Ecological Modernisation Theory." In *The Ecological Modernisation Reader: Environmental Reform in Theory and Practice*, edited by Arthur P. J. Mol, David A. Sonnenfeld, and Gert Spaargaren, 17–27. Abingdon, UK: Routledge, 2009.

Mol, Arthur P. J., G. Spaargaren, and D. A. Sonnenfeld. "Ecological Modernisation Theory: Where Do We Stand?" *Ökologische Modernisierung: Zur Geschichte Und Gegenwart Eines Konzepts in Umweltpolitik Und Sozialwissenschaften*. Frankfurt: Campus Verlag, 2014.

Moore, Susan A. "Defining 'Successful' Environmental Dispute Resolution: Case Studies from Public Land Planning in the United States and Australia." *Environmental Impact Assessment Review* 16, no. 3 (May 1996): 151–169.

Morehouse, Catherine. "NERA Counters Broad Opposition to FERC Net Metering Petition, Reveals Utility-Linked Member." *Utility Dive*, July, 2 2020. Accessed July 3, 2020. https://www.utilitydive.com/news/nera-counters-broad-opposition -to-ferc-net-metering-petition-reveals-1-uti.

Morris, Craig, and Arne Jungjohann. *Energy Democracy: Germany's Energiewende to Renewables*. London: Palgrave Macmillan, 2016.

Morrison, Peter A., and Judith P. Wheeler. *Rural Renaissance in America? The Revival of Population Growth in Remote Areas*. Washington, DC: Population Research Bureau, October 1976.

Morse, Cheryl E., Allan M. Strong, V. Ernesto Mendez, Sarah T. Lovell, Austin R. Troy, and William B. Morris. "Performing a New England Landscape: Viewing, Engaging, and Belonging." *Journal of Rural Studies* 36 (October 1, 2014): 226–236.

Mulvaney, Dustin. *Solar Power: Innovation, Sustainability, and Environmental Justice*. Berkeley: University of California Press, 2019.

Museum of Science and Industry. "Einstein May Outrank Britney Spears but Survey Shows Science Education Needs Help in United States: Data Gives New Look at the Opinions and Thoughts of Americans on the Critical Topic of Science." Press release, Museum of Science and Industry, Chicago, March 20, 2008.

Nadasdy, Paul. *Hunters and Bureaucrats: Power, Knowledge, and Aboriginal-State Relations in the Southwest Yukon*. Vancouver: University of British Columbia Press, 2004.

The New York Times. "Donald Trump's New York Times Interview: Full Transcript." November 23, 2016.

Newell, Peter, and Dustin Mulvaney. "The Political Economy of the 'Just Transition.'" *Geographical Journal* 179, no. 2 (2013): 132–140.

Nicholson-Cole, Sophie. "'Fear Won't Do It': Visual and Iconic Representations." *Science Communication* 30, no. 3 (2009): 355–379.

Nixon, Amy A. "Sheffield Wind Study: Turbines Too Loud." *Caledonian Record*, October 10, 2018.

Noble, Kim. "Wind Won't Be Our Savior." *Brattleboro Reformer*, August 5, 2006.

Ohanian, Hans. "Carbon Myopia in Montpelier." *VTDigger*, February 7, 2020.

Olson, Bradley. "Big Oil Companies Reap Windfall from Ethanol Rules." *Wall Street Journal*, October 27, 2016.

Olson-Hazboun, Shawn K. "'Why Are We Being Punished and They Are Being Rewarded?' Views on Renewable Energy in Fossil Fuels-Based Communities of the U.S. West." *Extractive Industries and Society* 5, no. 3 (July 1, 2018): 366–374.

"Opinion: Green Love Is Blind." *Wall Street Journal*, February 13, 2015.

Orsted. "Wind Power: Limitless Supply, Limitless Opportunity." Accessed September 11, 2018. https://orsted.com/en/Our-business/Wind-Power.

O'Shaughnessy, Eric, Jenny Heeter, Jeff Cook, and Christina Volpi. *Status and Trends in the U.S. Voluntary Green Power Market (2016 Data).* US Department of Energy, Office of Energy Efficiency & Renewable Energy, National Renewable Energy Laboratory, October 2017.

Pandey, Sudhanshu, Ritesh Gautam, Sander Houweling, Hugo Denier Van Der Gon, Pankaj Sadavarte, Tobias Borsdorff, Otto Hasekamp, Jochen Landgraf, Paul Tol, Tim van Kempen, Ruud Hoogeveen, Richard van Hees, Steven P. Hamburg, Joannes D. Maasakkers, and Ilse Aben. "Satellite Observations Reveal Extreme Methane Leakage from a Natural Gas Well Blowout." *Proceedings of the National Academy of Sciences of the United States of America* 116, no. 52 (December 26, 2019): 26376–26381.

Pazniokas, Mark. "Governors Want Sunlight on the Secretive ISO New England." *CT Mirror*, October 15, 2020.

Park, Lisa Sun-Hee, and David N. Pellow. *The Slums of Aspen: Immigrants vs. the Environment in America's Eden.* New York: New York University Press, 2013.

Parker, Bruce. "Public Service Commissioner: Vermont's Green Energy Plan Not about Global Warming." *Vermont Watchdog*, October 23, 2015.

Parkins, John R., Thomas Beckley, Louise Comeau, Richard C. Stedman, Curtis L. Rollins, and Anna Kessler. "Can Distrust Enhance Public Engagement? Insights from a National Survey on Energy Issues in Canada." *Society & Natural Resources* 30, no. 8 (August 3, 2017): 934–948.

Pepermans, Yves, and Ilse Loots. "Wind Farm Struggles in Flanders Fields: A Sociological Perspective." *Energy Policy* 59 (2013): 321–328.

Phadke, Roopali. "Public Deliberation and the Geographies of Wind Justice." *Science as Culture* 22, no. 2 (June 2013): 247–255.

Podobnik, Bruce. *Global Energy Shifts: Fostering Sustainability in a Turbulent Age.* Philadelphia: Temple University Press, 2006.

Polhamus, Mike. "Program to Promote Renewable Energy Gets New Scrutiny." *VT Digger*, January 2, 2018.

Polhamus, Mike. "Turbine Foes Seize on Window Issue in Temporary Sound Rules." *VT Digger*, August 4, 2016.

Post, Bruce, Curt McCormack, Charles W. Johnson, Kevin Jones (Petitioners), and Environmental and Natural Resources Law Clinic, Vermont Law School. "Petition to Investigate Deceptive Trade Practices of Green Mountain Power Company in the Marketing of Renewable Energy to Vermont Consumers." Filed September 16, 2014, Federal Trade Commission.

Prasanna, Ashreeta, Jasmin Mahmoodi, Tobias Brosch, and Martin K. Patel. "Recent Experiences with Tariffs for Saving Electricity in Households." *Energy Policy* 115 (April 1, 2018): 514–522.

"PUC Chairman Took Equity Stake in Wind Company." *Pine Tree Watch*, May 6, 2010.

Reicher, Dan. "Discuss Wind Power Based on Facts." *Burlington Free Press*, February 1, 2004.

Righter, Robert W. *Wind Energy in America: A History*. Norman: University of Oklahoma Press, 1996.

Ring, Wilson. "Vermont Wind-Turbine Noise Rules Displease Everyone." *Portland Press Herald*, November 13, 2017.

Rohracher, Harald. "Energy Systems in Transition: Contributions from Social Sciences." *International Journal of Environmental Technology and Management* 9, no. 2–3 (2008): 144–161.

———. "Managing the Technological Transition to Sustainable Construction of Buildings: A Socio-technical Perspective." *Technology Analysis & Strategic Management* 13, no. 1 (March 2001): 137–150.

Rooney, David. "Knowledge, Economy, Technology and Society: The Politics of Discourse." *Telematics and Informatics* 22, no. 4 (2005): 405–422.

Rubin, G James, Miriam Burns, and Simon Wessely. "Possible Psychological Mechanisms for 'Wind Turbine Syndrome': On the Windmills of Your Mind." *Noise & Health* 16, no. 69 (January 1, 2014): 116–122.

Sabourin, Clément. "In Quebec, Canada's Newest Hydroelectric Dams Nearly Ready." *Phys.org*, November 21, 2018.

Schalit, Naomi. "LD 1750: A Study in How Special Interests Get Their Way in the Maine Legislature." *Pine Tree Watch*, November 2015.

Schnaiberg, Allan. *The Environment: From Surplus to Scarcity*. Oxford: Oxford University Press, 1980.

Schreurs, Miranda, and Dörte Ohlhorst. "NIMBY and YIMBY." In *NIMBY Is Beautiful: Cases of Local Activism and Environmental Innovation around the World*, 1st ed., edited by Carol Hager and Mary Alice Haddad, 60–86. New York: Berghan Books, 2015.

Semega, Jessica, Melissa Kollar, John Creamer, and Abinash Mohanty. *Income and Poverty in the United States: 2018*. US Census Bureau, report no. P60-266, September 2019.

Sharkey, Betsy. "Review: 'Switch' Sticks to Energy Details." *Los Angeles Times*, November 8, 2012.

Shemkus, Sarah. "Maine Transmission Line Entangled by Conflicting Claims on Emissions." *Energy News Network*, November 2018.

Sherman, Michael, Gene Sessions, and P. Jeffrey Potash. *Freedom and Unity: A History of Vermont*. Barre: Vermont Historical Society, 2004.

Shove, Elizabeth. "Efficiency and Consumption: Technology and Practice." *Energy & Environment* 15, no. 6 (2004): 1053–1065.

Shultz, Jim. "The Renewable Energy Rebels." *New York Review of Books*, December 3, 2020.

Slocum, Rachel. "Polar Bears and Energy-Efficient Lightbulbs: Strategies to Bring Climate Change Home." *Environment and Planning D: Society and Space* 22, no. 3 (2004): 413–438.

Smith, Darren P., and Maureen H. McDonough. "Beyond Public Participation: Fairness in Natural Resource Decision Making." *Society & Natural Resources* 14, no. 3 (March 2001): 239–249.

Smith, Linda Tuhiwai. *Decolonizing Methodologies: Research and Indigenous Peoples*. London: Zed Books, 2013.

Sovacool, Benjamin K., Alex Gilbert, and Brian Thomson. "Innovations in Energy and Climate Policy: Lessons from Vermont." *Pace Environmental Law Review* 31, no. 3 (2014): 651–704.

Sovacool, Benjamin K., and Michael H. Dworkin. "Energy Justice: Conceptual Insights and Practical Applications." *Applied Energy* 142 (2015): 435–444.

Sovacool, Benjamin K., Mario Alberto Munoz Perea, Alfredo Villa Matamoros, and Peter Enevoldsen. "Valuing the Manufacturing Externalities of Wind Energy: Assessing the Environmental Profit and Loss of Wind Turbines in Northern Europe." *Wind Energy* 19, no. 9 (September 1, 2016): 1623–1647.

Stein, Andrew. "Shumlin Administration Forms Commission to Assess Siting Process for Industrial Wind and Other Energy Generation Projects." *VT Digger*, October 3, 2012.

Stirling, Andy. "Multicriteria Diversity Analysis: A Novel Heuristic Framework for Appraising Energy Portfolios." *Energy Policy* 38, no. 4 (2010): 1622–1634.

Stoll, Christian, Lena Klaaßen, and Ulrich Gallersdö. "The Carbon Footprint of Bitcoin." *Joule*, 2019, 1–15.

Szasz, Andrew. *Shopping Our Way to Safety: How We Changed from Protecting the Environment to Protecting Ourselves*. Minneapolis: University of Minnesota Press, 2007.

Therrien, Jim. "Teflon Town." *VT Digger*, September 1, 2017.

Toke, David, Sylvia Breukers, and Maarten Wolsink. "Wind Power Deployment Outcomes: How Can We Account for the Differences?" *Renewable and Sustainable Energy Reviews* 12, no. 4 (2008): 1129–1147.

Toke, David. *Ecological Modernisation and Renewable Energy*. London: Palgrave Macmillan, 2011.

Totten, Shay. "Tilting at Turbines." *Seven Days*, July 27, 2011.

Treffeisen, Beth. "Bourne Health Board Declares Plymouth Turbines a Nuisance." *Cape Cod Times*, October 25, 2018.

Trucost. *Plastics and Sustainability*. Trucost ESG Analysis, July 27, 2016.

Trust for Public Land. *New Hampshire's Return on Investment in Land Conservation*. June 2014.

Turkel, Tux. "Mainers Full of Gusto for Wind Power, Survey Finds." *Portland Press Herald*, June 29, 2010.

United States Geological Survey National Minerals Information Center. "Rare Earths." 2020.

Unruh, Gregory C. "Understanding Carbon Lock-In." *Energy Policy* 28, no. 12 (October 2000): 817–830.

Van der Horst, Dan. "NIMBY or Not? Exploring the Relevance of Location and the Politics of Voiced Opinions in Renewable Energy Siting Controversies." *Energy Policy* 35, no. 5 (2007): 2705–2714.

Vanderbeck, Robert M. "Vermont and the Imaginative Geographies of American Whiteness." *Annals of the Association of American Geographers* 96, no. 3 (2006): 641–659.

Varoufakis, Yanis. *Talking to My Daughter about the Economy: A Brief History of Capitalism*. London: Bodley Head, 2017.

Vasi, Ion Bogdan. *Winds of Change: The Environmental Movement and the Global Development of the Wind Energy Industry*. Oxford: Oxford University Press, 2011.

Verbong, Geert, and Frank Geels. "Future Electricity Systems." In *Governing the*

Energy Transition. Reality, Illusion or Necessity, edited by Geert Verbong and Derk Loorbach, 1–23. Abingdon, UK: Routledge, 2012.

———. "The Ongoing Energy Transition: Lessons from a Socio-technical, Multi-level Analysis of the Dutch Electricity System (1960–2004)." *Energy Policy* 35, no. 2 (2007): 1025–1037.

Vermont Department of Public Service. *2020 Annual Energy Report: A summary of progress made toward the goals of Vermont's Comprehensive Energy Plan.* 2020.

Vermont Commission on Country Life. "Rural Vermont: A Program for the Future." 1931. Accessed October 1, 2015. www.vbsr.org/about-vbsr.

Vermonters for a Clean Environment. "Understanding Vermont's Energy Policies Reaching the Goal of 90% Renewable by 2050." Accessed August 25, 2018. http://www.vce.org/VCE_WhitePaper_UnderstandingVermontEnergyPolicies _12March2018.pdf.

Way, Dan E. "Duke Energy Application Points Finger at Solar for Increased Pollution." *North State Journal*, August 14, 2019.

Weis, Elizabeth. "As Climate Change Threatens Earth, US to Open Nearly 200 Power Plants." *USA Today*, September 9, 2019.

Wertz, Joe. "Reading, Writing and Fracking? What the Oil Industry Teaches Oklahoma Students." National Public Radio, July 11, 2017.

Westervelt, Amy. "Drilled." *Critical Frequency*, February 3, 2020.

Wexler, Mark N. "A Sociological Framing of the NIMBY (Not-In-My-Backyard) Syndrome." *International Review of Modern Sociology* 26, no. 1 (Spring 1996): 91–110.

"Wind Turbines: Do Property Values Fall?" *St. Albans Messenger*, August 17, 2015.

Whitfield, Harvey Amani. *The Problem of Slavery in Early Vermont, 1777–1810.* Barre: Vermont Historical Society, 2014.

Whitlock, Robin. "Just Plain Wrong: The False Claims, Errors and Outdated Information That Is *Planet of the Humans*." *Renewable Energy Magazine*, May 13, 2020. Accessed June 29, 2020. https://www.renewableenergymagazine.com /panorama/just-plain-wrong-the-false-claims-errors-20200513.

Wilson, Elizabeth J., and Jennie C. Stephens. "Wind Deployment in the United States: States, Resources, Policy, and Discourse." *Environmental Science and Technology* 43, no. 24 (2009): 9063–9070.

Wolsink, Maarten. "Entanglement of Interests and Motives: Assumptions behind the NIMBY Theory on Facility Siting." *Urban Studies* 31, no. 6 (1994): 851–866.

Woods, Michael. "Deconstructing Rural Protest: The Emergence of a New Social Movement." *Journal of Rural Studies* 19, no. 3 (2003): 309–325.

Woolf, Art. "Fossil Fuel Consumption: Vermonters Driving More Miles in Their Cars." *Burlington Free Press*, August 30, 2018.

Worland, Justin. "Why Fossil Fuel Companies Are Reckoning with Climate Change." *Time*, January 2020.

Wright, Christopher, and Daniel Nyberg. *Climate Change, Capitalism, and Corporations.* Cambridge: Cambridge University Press, 2015.

Wüstenhagen, Rolf, Maarten Wolsink, and Mary Jean Burer. "Social Acceptance of Renewable Energy Innovation: An Introduction to the Concept." *Energy Policy* 35, no. 5 (May 2007): 2683–2691.

York, Richard, and Julius Alexander McGee. "Understanding the Jevons Paradox." *Environmental Sociology* 2, no. 1 (2016), 77–87.

York, Richard, and Shannon Elizabeth Bell. "Energy Transitions or Additions?: Why a Transition from Fossil Fuels Requires More than the Growth of Renewable Energy." *Energy Research & Social Science* 51 (2019): 40–43.

Zehner, Ozzie. "Conjuring Clean Energy: Exposing Green Assumptions in Media and Academia." *Foresight* 16, no. 6 (2014): 567–585.

———. *Green Illusions: The Dirty Secrets of Clean Energy and the Future of Environmentalism*. Lincoln: University of Nebraska Press, 2012.

———. "Unintended Consequences" In *Green Technology*, edited by Paul Robbins, Dustin Mulvaney and J. Geoffrey Golson, 427–432. London: Sage, 2011.

Index

Page numbers in *italics* represent figures and table.

Vermont: Act 56, 110, 183–184; Act 174, 38, 243; Act 250, 62; Agency of Natural Resources (ANR), 124, 127, 186, 198; banned billboards, 168, 253n15; carbon reduction policy, 21, 82; carbon tax, 168–169; Clean Energy Development Fund (CEDF), 121, 182, 259n6; Climate Council, 243; Commission on Country Life, 51; communal eco-identity, 168; conservation efforts, 88, 182; Department of Public Service, 85, 93, 122, 124; Department of Transportation, 67; ecological modernization theory and, 183–184, 187, 196; electricity comparison, 25, 182; electricity prices, 100; energy consumption linked to land use, 178–179, 191; Energy Efficiency and Conservation Block Grant (EECBG) program, 259n6; energy policy structure, 93; "green jobs," 117–118, 255n30; grid-scale projects, 28, 45–46; hydropower routes through, 58; impact of Tropical Storm Irene, 81; land preservation and conservation, 253n15; longer-range renewable electricity targets policy, 21; map of existing installations, 27; net-metering policy, 114; politicians' move into private sector, 121–122, 123; politics of reflexivity, 196–199; portrayed as Shangri-La in *Saturday Night Live*, 223–224; Public Service Board, 86, 182, 204; Public Utility Council (PUC), 27, 39, 122, 125; racism, 224–225; rate-setting methodologies, 126–127; REC policy, 33–34, 79, 85, 92–94, 104–110, 183–184, 226, 242–243; recycling programs, 253n15; regulators' close relationship with energy companies, 125–128; residential solar electric conversion, 183; secret energy planning, 127–128; siting law, 212; State Energy Program (SEP), 259n6; tourist industry, 34; tying into grid, 78; urban versus rural equity/prosperity, 34; utility companies, 56; vacation destination, 50; wind energy, 24, 27–28, 251n1; wind energy policies, 26, 242, 255n10 (*see also* Sustainably Priced Energy for Economic Development). *See also* electricity; Northern New England; nuclear power; wind energy

Vermont Electric Power Producers (VEPP), 93
Vermont Energy Investment Corporation (VEIC), 182
Vermont Public Interest Research Group (VPIRG), 122
Vermont Watchdog, 86
Vermont "way," 45, 46, 67
victim/victimization, 44–45, 151, 210–213
Vinalhaven, Maine, 190
VTDigger, 15–16, 145–146, 197, 247

Wall Street Journal, 123, 216, 239
Washington Post, 163
Washington State, 41–42
water contamination, 196–199, 200, 206, 227
weather patterns, 82
Weeks Act, 50
Westervelt, Amy, 163
West Virginia Marcellus Shale, 24, 143
white environmentalism, 219–224
White Mountain National Forest, 50, 58, 63, 215
white-nose syndrome, 65. *See also* wind turbine: interference with wildlife
wildlife. *See* wind turbine: interference with wildlife
wind energy: advancements in, 1; alliances, forms of, 118–119; carbon displacement, 31–32, 86, 136, 171; coalitions, 119–120; contribution to energy independence, 120; cost effectiveness of—project, 32; cost of litigation, 79; deregulation, 98–102; developing world's demand for, 141; development and political parties, 22–23; ecological benefits of, 171; ecological modernization theory and, 190–192; employment, 117–119, 120, 177; federal subsidies, 24, 27–28, 32, 77, 95, 102–104, 132–133; financial and environmental transparency, 31, 80–87, 88, 229–232; financing, 102–104; impacts on society, 80–87; incentivized by RECs, 108–110; injustice and, 208–226, 228 (*see also* social justice); justification of projects, 79; land swaps, 33; local involvement in, 202–206, *203*; management risk and uncertainty, 70–89; manipulation of uncertainty,

76–87; markets, 104–110, 116–120; mutual dependence and close alliance, 118–119; neoliberalism and national wind electricity landscape, 94–110; overestimation of productivity, 78, 85; partial alternative to fossil fuels, 31–32, 83, 153; place attachment and, 59, 61–63; policies, 25–26, 31, 34–40, 102–104; political opportunity structures for, 118; power density, 84–85; preferred renewable electricity option, 24; price stability, 120; profitability of, 170; rate setting, 77–78; ridgeline wind projects, 116, 135, 188–189, 244; risk and—, 76–87; sense of urgency to develop, 80–81; social scientists' approach to, 10–11; tax credits/incentives, 32, 77–78, 103, 255n16; unpopularity of, 23, 49; use in mitigating climate change, 81. *See also* energy transition; social issues and energy; wind turbine wind opponents/opposition: arguments about, 29; categorization of, 41–42; cohesive messaging, 40–41; described, 28–44; excluded from decision-making process, 55–56, 57, 58–59, 136; oversimplification of, 10–11; price volatility through litigation, 79; procedural arguments, 34–40, 45; public relations approach, 41; on regulators' close relationship with energy companies, 125–128; on selection of hosting communities, 209; technical arguments, 31–34, 45; use of term, 10. *See also* NIMBY (not in my backyard) phenomenon wind resisters. *See* wind opponents/opposition wind turbine: aesthetics debate, 30, 33, 44; as attractive investments, 102–104; backup fuels, 10, 144; blinking red lights, 3, 29; climate change and, 144; construction subsidies, 77; commoditizing—as financial instrument, 97, 102–104; criticisms related to mechanics and proximity, 28, 29–30; decommissioning,

removal, disposal, 30, 78, 80, 217; development, 77; drone technology monitoring, 173–174; ecological cost of, 32–33, 49, 70, 80–87, 88, 166, 171–172; ecological modernization theory and, 194–196; effectiveness for climate mitigation, 83–84; effect on home values, 11, 30, 252n13; financing models, 10; fire hazard, 29, 71, 72; flickering shadows, 29; gearboxes, 85; grid-scale versus smaller-scale, 116, 252n2; hosting communities, 16; ice from, 29; impact on weather patterns, 82; interference with wildlife, 28, 30, 33, 50, 65, 88, 229–230; life span, 85; lightning strikes, 154; in Lowell, Vermont, 2–7, 4; map of existing installations, 27; mining and mineral processing used in making, 217; moratorium, 204, 242; oil and coolant leakages, 29, 70; onshore vs. offshore turbine, 171–172; opposition to, 2–7; ownership arrangements, 10, 11, 30; payments for hosting, 37; permanent employment, 34; ridgeline, 6, 31–32, 45, 111–112, 190–192, 217; sickness from, 30, 252n11; site preparation costs, 84–85; siting in rural areas, 33, 34, 45, 47, 55–56, 57, 58–59, 136, 202–206, 208–209, 210–213; size of, 29, 46; sound from, 28, 30, 88, 166, 209, 242, 252n11; symbolism of, 46, 47, 165, 166–172, 176, 229–232; technology, 10, 170; tying to grid, 78. *See also* wind energy wind turbine syndrome, 252n11 windwatch.org, 29 winterization programs, 132 Woods, Michael, 61 World Bank, 216 World Health Organization, 76

Yieldcos, 103–104

Zehner, Ozzie, 161–163, 167

About the Author

SHAUN A. GOLDING is an assistant professor of sociology at Kenyon College. He is a graduate of Bowdoin College and the University of Wisconsin–Madison, and he recently completed a Fulbright fellowship in Norway.

Available titles in the Nature, Society, and Culture series

Printed in the United States
by Baker & Taylor Publisher Services